SpringerBriefs in Computer Science

More information about this series at http://www.springer.com/series/10028

M.N. Murty · Rashmi Raghava

Support Vector Machines and Perceptrons

Learning, Optimization, Classification, and Application to Social Networks

 Springer

M.N. Murty
Department of Computer Science
  and Automation
Indian Institute of Science
Bangalore, Karnataka
India

Rashmi Raghava
IBM India Private Limited
Bangalore, Karnataka
India

ISSN 2191-5768   ISSN 2191-5776 (electronic)
SpringerBriefs in Computer Science
ISBN 978-3-319-41062-3   ISBN 978-3-319-41063-0 (eBook)
DOI 10.1007/978-3-319-41063-0

Library of Congress Control Number: 2016943387

Printed on acid-free paper

This Springer imprint is published by Springer Nature
The registered company is Springer International Publishing AG Switzerland

Preface

Overview

Support Vector Machines (SVMs) have been widely used in *Classification, Clustering and Regression*. In this book, we deal primarily with classification. Classifiers can be either *linear* or *nonlinear*. The linear classifiers typically are learnt based on a *linear discriminant function* that separates the feature space into two half-spaces, where one half-space corresponds to one of the two classes and the other half-space corresponds to the remaining class. So, these half-space classifiers are ideally suited to solve *binary classification* or two-class classification problems. There are a variety of schemes to build multiclass classifiers based on combinations of several binary classifiers.

Linear discriminant functions are characterized by a *weight vector* and a *threshold* weight that is a scalar. These two are learnt from the training data. Once these entities are obtained we can use them to classify patterns into any one of the two classes. It is possible to extend the notion of linear discriminant functions (LDFs) to deal with even nonlinearly separable data with the help of a suitable mapping of the data points from the low-dimensional *input* space to a possibly higher dimensional *feature space*.

Perceptron is an early classifier that successfully dealt with linearly separable classes. Perceptron could be viewed as the simplest form of *artificial neural network*. An excellent theory to characterize parallel and distributed computing was put forth by *Misky and Papert* in the form of a book on perceptrons. They use logic, geometry, and group theory to provide a computational framework for perceptrons. This can be used to show that any computable function can be characterized as a linear discriminant function possibly in a high-dimensional space based on *minterms* corresponding to the input Boolean variables. However, for some types of problems one needs to use all the minterms which correspond to using an exponential number of minterms that could be realized from the primitive variables.

SVMs have revolutionized the research in the areas of *machine learning* and *pattern recognition*, specifically classification, so much that for a period of more than two decades they are used as state-of-the-art classifiers. Two distinct properties of SVMs are:

1. The problem of learning the LDF corresponding to SVM is posed as a convex optimization problem. This is based on the intuition that the hyperplane separating the two classes is learnt so that it corresponds to maximizing the *margin* or some kind of separation between the two classes. So, they are also called as *maximum-margin classifiers*.
2. Another important notion associated with SVMs is the *kernel trick* which permits us to perform all the computations in the low-dimensional input space rather than in a higher dimensional feature space.

These two ideas become so popular that the first one lead to the increase of interest in the area of *convex optimization*, whereas the second idea was exploited to deal with a variety of other classifiers and clustering algorithms using an appropriate kernel/similarity function.

The current popularity of SVMs can be attributed to excellent and popular software packages like *LIBSVM*. Even though SVMs can be used in nonlinear classification scenarios based on the kernel trick, the linear SVMs are more popular in the real-world applications that are high-dimensional. Further learning the parameters could be time-consuming. There is a renewal of energy, in the recent times, to examine other linear classifiers like perceptrons. Keeping this in mind, we have dealt with both perceptron and SVM classifiers in this book.

Audience

This book is intended for senior undergraduate and graduate students and researchers working in *machine learning, data mining, and pattern recognition*. Even though SVMs and perceptrons are popular, people find it difficult to understand the underlying theory. We present material in this book so that it is accessible to a wide variety of readers with some basic exposure to undergraduate level mathematics. The presentation is intentionally made simpler to make the reader feel comfortable.

Organization

This book is organized as follows:

1. *Literature and Background*: Chapter 1 presents literature and state-of-the-art techniques in SVM-based classification. Further, we also discuss relevant background required for pattern classification. We define some of the important terms that are used in the rest of the book. Some of the concepts are explained with the help of easy to understand examples.

2. *Linear Discriminant Function*: In Chap. 2 we introduce the notion of *linear discriminant function* that forms the basis for the linear classifiers described in the text. The role of *weight vector W* and the *threshold b* are explained in describing linear classifiers. We also describe other linear classifiers including the *minimal distance classifier* and the *Naïve* Bayes classifier. It also explains how nonlinear discriminant functions could be viewed as linear discriminant functions in higher dimensional spaces.

3. *Perceptron*: In Chap. 3 we describe perceptron and how it can be used for classification. We deal with *perceptron learning algorithm* and explain how it can be used to learn Boolean functions. We provide a simple proof to show how the algorithm converges. We explain the notion of *order of a perceptron* that has bearing on the computational complexity. We illustrate it on two different classification datasets.

4. *Linear SVM*: In this Chap. 4, we start with the similarity between SVM and perceptron as both of them are used for *linear classification*. We discuss the difference between them in terms of the form of computation of w, the optimization problem underlying each, and the *kernel trick*. We introduce the linear SVM which possibly is the most popular classifier in machine learning. We introduce the notion of *maximum margin* and the geometric and semantic interpretation of the same. We explain how a binary classifier could be used in building a multiclass classifier. We provide experimental results on two datasets.

5. *Kernel Based SVM*: In Chap. 5, we discuss the notion of *kernel* or similarity function. We discuss how the optimization problem changes when the classes are not linearly separable or when there are some data points on the margin. We explain in simple terms the *kernel trick* and explain how it is used in classification. We illustrate using two practical datasets.

6. *Application to Social Networks*: In Chap. 6 we consider *social networks*. Specifically, issues related to representation of social networks using *graphs*; these graphs are in turn represented as matrices or lists. We consider the problem of community detection in social networks and *link prediction*. We examine several existing schemes for link prediction including the one based on SVM classifier. We illustrate its working based on some network datasets.

7. *Conclusion*: We conclude in Chap. 7 and also present potential future directions.

Bangalore, India   M.N. Murty
                                                                                              Rashmi Raghava

Contents

Acronyms

CC	Clustering Coefficient
DTC	Decision Tree Classifier
KKT	Karush Kuhn Tucker
KNNC	K-Nearest Neighbor Classifier
LDF	Linear Discriminant Function
MDC	Minimal Distance Classifier
NBC	Naïve Bayes Classifier
NNC	Nearest Neighbor Classifier
SVM	Support Vector Machine

Chapter 1
Introduction

Abstract Support vector machines (SVMs) have been successfully used in a variety of data mining and machine learning applications. One of the most popular applications is pattern classification. SVMs are so well-known to the pattern classification community that by default, researchers in this area use them as baseline classifiers to establish the superiority of the classifier proposed by them. In this chapter, we introduce some of the important terms associated with support vector machines and a brief history of their evolution.

Keywords Classification · Representation · Proximity function · Classifiers

Support Vector Machine (SVM) [1, 2, 5, 6] is easily the most popular tool for pattern classification; by *classification* we mean the process of assigning a class label to an unlabeled pattern using a set of labeled patterns. In this chapter, we introduce the notion of classification and classifiers. First we explain the related concepts/terms; for each term we provide a working definition, any philosophical characterization, if necessary and the notation.

1.1 Terminology

First we describe the terms that are important and used in the rest of the book.

1.1.1 What Is a Pattern?

A pattern is either a *physical object* or an *abstract notion*.

We need such a definition because in most of the practical applications, we encounter situations where we have to classify physical objects like *humans, chairs, and a variety of other man-made objects*. Further, there could be applications where classification of abstract notions like *style of writing, style of talking, style of walking, signature, speech, iris, finger-prints of humans* could form an important part of the application.

M.N. Murty and R. Raghava, *Support Vector Machines and Perceptrons*,
SpringerBriefs in Computer Science, DOI 10.1007/978-3-319-41063-0_1

1.1.2   Why Pattern Representation?

In most machine-based pattern classification applications, patterns cannot be directly stored on the machine. For example, in order to discriminate humans from chairs, it is not possible to store either a human or a chair directly on the machine. We need to represent such patterns in a form amenable for machine processing and store the representation on the machine.

1.1.3   What Is Pattern Representation?

Pattern representation is the process of generating an abstraction of the pattern which could be stored on the machine.

For example, it is possible to represent chairs and humans based on their *height* or in terms of their *weight* or *both height and weight*. So patterns are typically represented using some scheme and the resulting representations are stored on the machine.

1.1.4   How to Represent Patterns?

Two popular schemes for pattern representation are:

1. **Vector Space** representation: Here, a pattern is represented as a *vector* or a *point* in a multidimensional space.
   For example $(1.2, 4.9)^t$ might represent a *chair* of *height* 1.2 m and *weight* 4.9 kg.
2. **Linguistic/Structural** representation: In this case, a pattern is represented as a sentence in a formal language.
   For   example, $(color = red \vee white) \wedge (make = leather) \wedge (shape = sphere) \wedge (dia = 7\,cm) \wedge (weight = 150\,g)$ might represent *cricket ball.*

We will consider only *vector representations* in this book.

1.1.5   Why Represent Patterns as Vectors?

Some of the important reasons for representing patterns as vectors are:

1. *vector space* representations are popular in pattern classification. Classifiers based on fuzzy sets, rough sets, statistical learning theory, decision tree classifiers all are typically used in conjunction with patterns represented as vectors.

2. Classifiers based on neural networks and support vector machines are inherently constrained to deal only with vectors of numbers.
3. Pattern recognition algorithms that are typically based on similarity/dissimilarity between pairs of patterns use metrics like *Euclidean distance*, and similarity functions like *cosine of the angle between vectors*; these proximity functions are ideally suited to deal with vectors of reals.

1.1.6 Notation

- **Pattern**: Even though pattern and its representation are different, it is convenient and customary to use *pattern* for both.
  The usage is made clear based on the context in which the term is used; on a machine, for pattern classification, a representation of the pattern is stored, not the pattern itself. In the following, we will be concerned only with pattern representation; however, we will call it pattern as is practiced.
  We use X to represent a pattern.
- **Collection of Patterns**: A collection of n patterns is represented by $\{X_1, X_2, \ldots, X_n\}$ where X_i denotes the ith pattern.
  We assume that each pattern is an l-dimensional vector.

$$\text{So,} X_i = (x_{i1}, x_{i2}, \ldots, x_{il}).$$

1.2 Proximity Function [1, 4]

The notion of proximity is typically used in classification. This is characterized by either a *distance function* or a *similarity function*

1.2.1 Distance Function

Distance between patterns X_i and X_j is denoted by $d(X_i, X_j)$ and the most popular distance measure is the *Euclidean distance* and it is given by

$$d(X_i, X_j) = \left(\sum_{k=1}^{l} (x_{ik} - x_{jk})^2 \right)^{\frac{1}{2}}$$

A pair of patterns are closer or similar if the distance between them is smaller.

Euclidean distance is a *metric* and so it satisfies, for any three patterns X_i, X_j, and X_k, the following properties:

1. $d(X_i, X_j) \geq 0$ *(Nonnegativity)*
2. $d(X_i, X_j) = d(X_j, X_i)$ *(Symmetry)*
   Symmetry is useful in reducing the storage requirements because it is sufficient to store either $d(X_i, X_j)$ or $d(X_j, X_i)$, both are not required.
3. $d(X_i, X_j) + d(X_j, X_k) \geq d(X_i, X_k)$ *(Triangle Inequality)*
   Triangle inequality is useful in reducing the computation time and also in establishing some useful bounds to simplify the analysis of several algorithms.

Even though metrics are useful in terms of computational requirements, they are not essential in *ranking* and *classification*.

For example, *squared Euclidean distance* is not a metric; however, it is as good as the Euclidean distance in both ranking and classification.

Example

Let $X = (1, 1)^t$, $X_1 = (1, 3)^t$, $X_2 = (4, 4)^t$, $X_3 = (2, 1)^t$

Note that $d(X, X_3) = 1 < d(X, X_1) = 2 < d(X, X_2) = 3\sqrt{2}$

Note that smaller the distance, nearer the pattern. So, the first three neighbors of X based on Euclidean distance are X_3, X_1, and X_2 in that order.

Similarly, the squared Euclidean distances are

$d(X, X_3)^2 = 1 < d(X, X_1)^2 = 4 < d(X, X_2)^2 = 9$. So, the first three neighbors of X based on squared distance are X_3, X_1, and X_2 in the same order again.

Consider two more patterns, $X_4 = (3, 3)^t$, and $X_5 = (5, 5)^t$. Note that the squared Euclidean distances are

$d(X, X_4)^2 = 8 = d(X_4, X_5)^2$; however, $d(X, X_5)^2 = 32 > d(X, X_4)^2 + d(X_4, X_5)^2$. So, triangle inequality is not satisfied by the squared euclidean distance.

1.2.2 Similarity Function

Cosine of the angle between vectors is the most popular similarity function. It is defined as follows:

$$\cos(X_i, X_j) = \frac{X_i^t X_j}{\| X_i \| \| X_j \|}$$

Here, the numerator characterizes the dot product or sum of the products of components and is given by

$$X_i^t X_j = \sum_{k=1}^{l} x_{ik} x_{jk}$$

Considering the patterns X, X_1, X_2, and X_3 seen in the previous example again we have

$$\cos(X, X_2) = 1 \; > \; \cos(X, X_3) = \frac{3}{\sqrt{10}} \; > \; \cos(X, X_1) = \frac{4}{\sqrt{20}} = \frac{2}{\sqrt{5}}.$$

So, the first three neighbors of X in the order of similarity are X_2, X_3, and X_1.

Note that X and X_2 are very similar using the cosine similarity as these two patterns have an angle of 0 degrees between them, even though they have different magnitudes. The magnitude is emphasized by the Euclidean distance; so X and X_2 are very dissimilar. This is exploited in high-dimensional applications like *text mining* and *information retrieval* where the cosine similarity is more popularly used. The reason may be explained as follows:

Consider a document d; let it be represented by X. Now consider a new document obtained by appending d to itself 3 times thus giving us $4X$ as the representation of the new document.

So, for example, if $X = (1, 1)^t$ then the new document is represented by $(4, 4)^t$. Note that the Cosine similarity between $(1, 1)^t$ and $(4, 4)^t$ is 1 as there is no difference between the two in terms of the semantic content.

However, in terms of Euclidean distance, $d((1, 1)^t, (4, 4)^t)$ is larger than $d((1, 1)^t, (2, 1)^t)$ whereas the Cosine similarity between $(1, 1)^t$ and $(2, 1)^t$ is smaller than that between X and $(4, 4)^t$.

1.2.3 Relation Between Dot Product and Cosine Similarity

Consider three patterns: $X_i = (1, 2)^t$, $X_j = (4, 2)^t$, $X_k = (2, 4)^t$. We give in the table the dot product and Cosine similarity values between all the possible pairs (Table 1.1).

Note that the *dot product* and *cosine similarity* are not linked monotonically. The dot product value is increasing from pair 1 to pair 3; however, it is not the case with the cosine similarity.

If the patterns are *normalized* to be unit norm vectors, then there is no difference between the dot product and cosine similarity. This is because

$$dot\,product(X_p, X_q) = X_p^t X_q = \frac{X_p^t X_q}{\| X_p \| \, \| X_q \|} = Cosine(X_p, X_q)$$

This equality holds because $\| X_p \| = \| X_q \| = 1$.

Table 1.1 Dot product and cosine similarity

Pair number	Pattern pair	Dot product	Cosine similarity
1	(X_i, X_j)	8	0.8
2	(X_i, X_k)	10	1
3	(X_j, X_k)	16	0.8

1.3   Classification [2, 4]

1.3.1   Class

A class is a collection/set of patterns where each pattern in the collection is associated with the same *class label*.

Consider a two-class problem where C_- is the *negative class* and C_+ is the *positive class*.

1.3.2   Representation of a Class

It is possible to characterize these two classes as follows. For some function g such that

$g : \mathbb{R}^l \to \mathbb{R}.$
$C_- = \{X | g(X) < 0\}$ and
$C_+ = \{X | g(X) > 0\}$

Note that the function g maps an $l - dimensional$ pattern X to a real number. We can think of more general functions, like for example the co-domain of g can be set of complex numbers, $\mathbb{C}$ instead of the set of reals, $\mathbb{R}$. However, the above definition, based on reals, is adequate for our study.

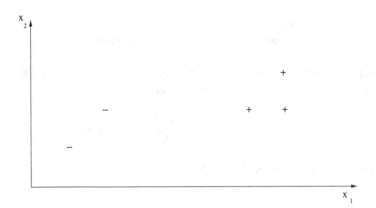

Fig. 1.1 An example dataset

1.3.3 Choice of G(X)

It is possible to choose the form of $g(X)$ in a variety of ways. We examine some of them next. We illustrate these choices using five two-dimensional pattern shown in Fig. 1.1. Note that we are considering the two classes to be represented as follows:

$$\mathbf{C}_- = \{(1, 1)^t, (2, 2)^t\}, \ \mathbf{C}_+ = \{(6, 2)^t, (7, 2)^t, (7, 3)^t\}.$$

1.4 Classifiers

1.4.1 Nearest Neighbor Classifier (NNC)

The nearest neighbor classifier obtains the nearest neighbor, from the training data, of the test pattern X. If the nearest neighbor is from C_- then it assigns X to C_-. Similarly, X is assigned to C_+ if the nearest neighbor of X is from class C_+.

Consider $g(X) = g_-(X) - g_+(X)$ for some $X \in \mathbb{R}^l$. Here,

$$g_-(X) = \min_{X_j \in C_-} d(X, X_j)$$

and

$$g_+(X) = \min_{X_j \in C_+} d(X, X_j)$$

for some distance function $d(-, -)$. We illustrate it with the example data shown in Fig. 1.1.

Let $X = (1, 2)^t$ and let $d(-, -)$ be the *squared euclidean distance*.

Note that $g_-(X) = 1$ and $g_+(X) = 25$

So, $g(X) = -24 < 0$, as a consequence, X is assigned to C_-.

If we consider $X = (5, 2)^t$, then $g(X) = 9 - 1 = 8 > 0$. So, X is assigned to C_+.

Note that the classifier based on $g(X)$ is the *NNC* for the two-class problem.

1.4.2 K-Nearest Neighbor Classifier (KNNC)

The *KNNC* obtains K neighbors of the test pattern X from the training data. If a majority of these K neighbors are from C_- then X is assigned to C_-. Otherwise, X is assigned to C_+.

In this case, $g(X) = g_+(X) - g_-(X)$ where $g_-(X) = K_-$ and $g_+(X) = K_+ = K - K_-$.

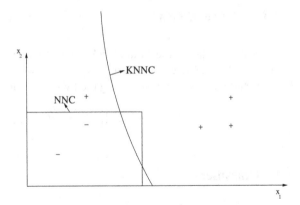

Fig. 1.2 Decision boundaries for NNC and KNNC

We obtain K-nearest neighbors of X from $C_- \bigcup C_+$. Let K_- (out of K) be the number of neighbors identified from C_- and the remaining $K_+ = K - K_-$ be the neighbors from C_+.

Note that $g(X) < 0$ if $K_+ < K_-$.

Now consider $X = (1, 2)^t$ and $K = 3$. The three neighbors are $(1, 1)^t$, $(2, 2)^t$, and $(6, 2)^t$.

Then $g_-(X) = K_- = 2$ and $g_+(X) = K_+ = 1$. So, $g(X) = 1 - 2 = -1 < 0$. Hence X is assigned to C_-.

Similarly, if we consider $X = (5, 2)^t$, then the three neighbors are $(6, 2)^t$, $(7, 2)^t$, and $(7, 3)^t$.

So, $g_-(X) = K_- = 0$ and $g_+(X) = K_+ = 3$. Here, $g(X) = 3 - 0 = 3 > 0$. So, X is assigned to C_+.

Note that the classifier based on $g(X)$ is *KNNC* corresponding to the two-class problem.

It is possible to observe that both *NNC* and *KNNC* can lead to nonlinear decision boundaries as shown in Fig. 1.2. Here, *NNC* gives a piecewise linear decision boundary and the KNNC gives a nonlinear decision boundary as depicted in the figure.

1.4.3 Minimum-Distance Classifier (MDC)

The working of MDC is as follows:

Let m_- and m_+ be the sample means of C_- and C_+ respectively. Assign the test pattern X to C_- if
$d(X, m_-) < d(X, m_+)$ else assign X to C_+.

Consider again $g(X) = g_-(X) - g_+(X)$ for some $X \in \mathbb{R}^l$. Here,

$$g_-(X) = d(X, m_-)$$

and

$$g_+(X) = d(X, m_+)$$

where $d(-, -)$ is some distance function and
Sample mean of points in $C_- =$

$$m_- = \frac{\sum_{X_j \in C_-} X_j}{|C_-|}$$

Sample mean of points in $C_+ =$

$$m_+ = \frac{\sum_{X_j \in C_+} X_j}{|C_+|}$$

We illustrate it with the example data shown in Fig. 1.1.

Note that $m_- = (1.5, 1.5)^t$ and $m_+ = (6.66, 2.33)^t$. So, if $X = (1, 2)^t$, then using the squared Euclidean distance for $d(-, -)$, we have $g_-(X) = 0.5$ and $g_+(X) = 32.2$; so, $g(X) = -31.7 < 0$. Hence, X is assigned to C_-.

If we consider $X = (5, 2)^t$, then $g(X) = 12.5 - 2.9 = 9.6 > 0$. Hence, X is assigned to C_+.

It is possible to show that the MDC is as good as the optimal classifier (Bayes classifier) if the two classes C_- and C_+ are normally distributed with $\mathcal{N}(\mu_i, \Sigma_i)$, $i = 1, 2$, where the covariance matrices Σ_1 and Σ_2 are such that $\Sigma_1 = \Sigma_2 = \sigma^2 I$, I being the Identity matrix and $\mu_1 = m_-$ and $\mu_2 = m_+$.

It is possible to show that the sample mean m_i converges to the true mean μ_i asymptotically or if the number of training patterns in each class is large in number.

1.4.4 Minimum Mahalanobis Distance Classifier

Let $g_-(X) = (X - \mu_-)^t \Sigma^{-1}(X - \mu_-)$ and
$g_+(X) = (X - \mu_+)^t \Sigma^{-1}(X - \mu_+)$

Note that $g_-(X)$ and $g_+(X)$ are squared Mahalanobis distances between X and μ_- and μ_+ respectively.

The minimum Mahalanobis distance classifier assigns X to C_- if $g_-(X) < g_+(X)$ else X is assigned to C_+.

If the data in each class is normally distributed and we are given that $\Sigma_1 = \Sigma_2 = \Sigma$ and no further structure on the *covariance matrices*, then the Mahalanobis distance classifier can be shown to be optimal. It is given by
$g(X) = g_-(X) - g_+(X)$ where
$g_-(X) = (X - \mu_-)^t \Sigma^{-1}(X - \mu_-)$ and
$g_+(X) = (X - \mu_+)^t \Sigma^{-1}(X - \mu_+)$

Note that $g_-(X)$ and $g_+(X)$ are the squared Mahalanobis distances between X and the respective classes.

Note that an estimate of Σ can be obtained by using all the five patterns and using the estimate for *Sigma* to be

$$\Sigma = \frac{1}{5} \sum_{i=1}^{l} (X_i - m)(X_i - m)^t]$$

where m is the mean of the five patterns and is given by $m = (4.6, 2)^t$.

Note that the estimated value for Σ is

$$\Sigma = \begin{bmatrix} 5.8 & -0.24 \\ -0.24 & 0.4 \end{bmatrix}$$

So, Σ^{-1} is given by

$$\Sigma^{-1} = \frac{4}{7} \begin{bmatrix} 0.4 & 0.24 \\ -0.24 & 5.8 \end{bmatrix}$$

If we choose $X = (1, 2)^t$ then $g(X) = 0.9 - 7.9 = -7 < 0$ and so X is assigned to C_- by using all the five patterns in the estimation of Σ.

Instead if we choose $X = (5, 2)^t$, then $g(X) = 3.6 - 0.4 = 3.2 > 0$ and so X is assigned to C_+.

1.4.5   Decision Tree Classifier: (DTC)

In the case of DTC, we find the best split based on the given features. The best feature is the one which separates the patterns belonging to the two classes so that each part is as *pure* as possible. Here, by purity we mean patterns are all from the same class. For example, consider the dataset shown in Fig. 1.3. Splitting on feature X_1 gives two parts. The right side part is from class C_+ (pure) and the left side part has more patterns from C_- with impurity in the form of one positive pattern. Splitting on X_2 may leave us with more impurity.

Again we have $g(X) = g_+(X) - g_-(X)$. Here, $g_+(X)$ and $g_-(X)$ are Boolean functions taking a value of either 1 or 0. Each leaf node in the decision tree is associated with one of the two class labels.

If there are m leaf nodes out of which m_- are associated with class C_- and remaining are positive, then $g_-(X)$ is a disjunction of m_- conjunctions and similarly $g_+(X)$ is a disjunction of $(m - m_-)$ conjunctions where each conjunction corresponds to a path from the root to a leaf.

In the data shown in Fig. 1.3.

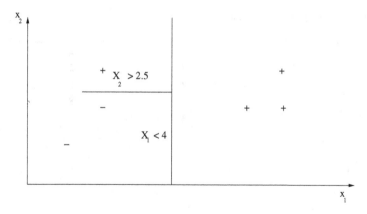

Fig. 1.3 An example dataset

Fig. 1.4 Decision tree for
the data

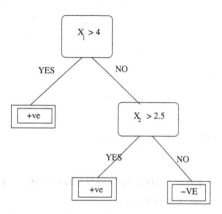

There are six patterns and the class labels for them are:

- **Negative Class:** $(1, 1)^t$, $(2, 2)^t$
- **Positive Class:** $(2, 3)^t$, $(6, 2)^t$, $(7, 2)^t$, $(7, 3)^t$

The corresponding decision tree is shown in Fig. 1.4. There are three leaf nodes in the tree; one is negative and two are positive. So, the corresponding $g_-(X)$ and $g_+(X)$ are:

- $g_-(X) = (x_1 \leq 4) \wedge (x_2 \leq 2.5)$ and
- $g_+(X) = (x_1 > 4) \vee (x_1 \leq 4) \wedge (x_2 > 2.5)$

If $X = (1, 2)^t$, then $g_-(X) = 1$ and $g_+(X) = 0$ (assuming that a Boolean function returns a value 0 when it is *FALSE* and a value 1 when it is *TRUE*. So, $g(X) = g_+(X) - g_-(X) = 0 - 1 = -1 < 0$; hence X is assigned to C_-.
If $X = (5, 2)^t$, then $g_-(X) = 0$ and $g_+(X) = 1$. So, $g(X) = 1$; hence X is assigned to C_+.

Table 1.2 Linear discriminant function

Pattern number	x_1	x_2	$g(X) = 2x_1 - 2x_2 - 2$
1	1	1	-2
2	2	2	-2
3	6	2	6
4	7	2	8
5	7	3	6

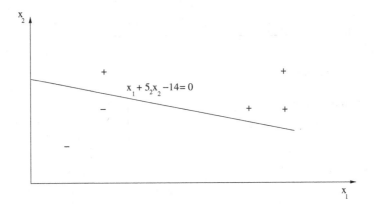

Fig. 1.5 Linear discriminant

1.4.6 Classification Based on a Linear Discriminant Function

Typically, we consider $g(X) = W^t X + w_0$ where W is an l-dimensional vector given by $W = (w_1, w_2, \ldots, w_l)^t$ and w_0 is a scalar. It is linear in both W and X.

In the case of the data shown in Fig. 1.1, let us consider $W = (2, -2)^t$ and $w_0 = -2$. The values of X and $g(X) = 2x_1 - 2x_2 - 2$ are shown in Table 1.2

Note that $g(X) < 0$ for $X \in C_-$ and $g(X) > 0$ for $X \in C_+$.

If we add to this set another pattern $(2, 3)^t (\in C_+)$ as shown in Fig. 1.3, then $g(X) = 2x_1 - 2x_2 - 2$ will not work. However, it is possible to show that $g(X) = x_1 + 5x_2 - 14$ classifies all the six patterns correctly as shown in Fig. 1.5.

We will discuss algorithms to obtain W and w_0 from the data in the later chapters.

1.4.7 Nonlinear Discriminant Function

Here $g(X)$ is nonlinear in X. For example, consider $g(X) = w_1 x_1^2 + w_2 x_2 + w_0$. For the example data in Fig. 1.1, we show the values in Table 1.3.

Table 1.3 Nonlinear discriminant function

Pattern number	x_1	x_2	$g(X) = 7x_1^2 - 16x_2 - 10$
1	1	1	−19
2	2	2	−14
3	6	2	210
4	7	2	301
5	7	3	285

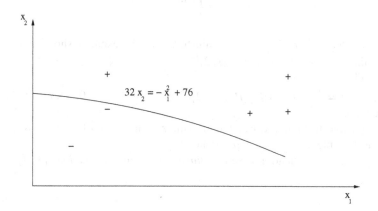

$$32\,x_2 = -x_1^2 + 76$$

Fig. 1.6 Nonlinear discriminant

Again we have $g(X) < 0$ for patterns in C_- and $g(X) > 0$ for patterns in C_+. Now consider the six patterns shown in Fig. 1.3. The function $7x_1^2 - 16x_2 - 10$ fails to classify the pattern $(2, 3)'$ correctly.

However, the function $g(X) = x_1^2 + 32x_2 - 76$ correctly classifies all the patterns as shown in Fig. 1.6. We will consider learning the nonlinear discriminant function later.

1.4.8 Naïve Bayes Classifier: (NBC)

NBC works as follows:
Assign X to C_- if $P(C_-|X) > P(C_+|X)$ else assign X to C_+.
Here, $g(X) = g_-(X) - g_+(X)$ where $g_-(X) = P(C_-|X)$ and $g_+(X) = P(C_+|X)$.
Using Bayes rule we have

$$P(C_-|X) = \frac{P(X|C_-)P(C_-)}{P(X)}$$
$$P(C_+|X) = \frac{P(X|C_+)P(C_+)}{P(X)}$$

Further, in NBC primarily we assume *class-conditional independence*. So, we get

$$P(C_-|X) = \prod_{i=1}^{l} P(x_i|C_-)P(C_-)$$

and

$$P(C_+|X) = \prod_{i=1}^{l} P(x_i|C_+)P(C_+)$$

We assume $P(C_-) = P(C_+) = \frac{1}{2}$. We illustrate with the example shown in Fig. 1.1.
Note that for $X = (1, 2)^t$, $P(C_-|(1, 2)^t) = \frac{1}{2}P(x_1 = 1|C_-)P(x_2 = 2|C_-)$
$= (\frac{1}{2})(\frac{1}{2})(\frac{1}{2}) = \frac{1}{8}$ and
$P(C_+|(1, 2)^t) = \frac{1}{2}P(x_1 = 1|C_+)P(x_2 = 2|C_+) = (\frac{1}{2})(0)(0) = 0.$ So, $(1, 2)^t$ is
assigned to C_-.

It is possible to view most of the classifiers dealing with binary classification
(two-class) problems using an appropriate $g(X)$.

We consider classification based on *linear discriminant functions* [3, 4] in this
book.

1.5 Summary

In this chapter, we have introduced the terms and notation that will be used in the
rest of the book. We stressed the importance of representing patterns and collections
of patterns. We described some of the popular distance and similarity functions that
are used in *machine learning*.

We introduced the notion of a discriminant function that could be useful in
abstracting classifiers. We have considered several popular classifiers and have shown
how they can all be abstracted using a suitable discriminant function in each case.
Specifically, we considered *NNC, KNNC, MDC, DTC, NBC*, and classification based
on *linear* and *nonlinear discriminant functions*.

References

1. Abe, S.: Support Vector Machines for Pattern Classification. Springer (2010)
2. Cristianini, N., Shawe-Taylor, J.: An Introduction to Support Vector Machines. Cambridge University Press (2000)
3. Minsky, M.L., Papert, S.: Perceptrons: An Introduction To Computational Geometry. MIT Press (1969)
4. Murphy, K.P.: Machine Learning—A Probabilistic Perspective. MIT Press (2012)
5. Vapnik, V.: The Nature of Statistical Learning Theory. Springer (2000)
6. Wang, L.: Support Vector Machines: Theory and Applications. Springer (2005)

Chapter 2
Linear Discriminant Function

Abstract Linear discriminant functions (LDFs) have been successfully used in pattern classification. Perceptrons and Support Vector Machines (SVMs) are two well-known members of the category of *linear discriminant functions* that have been popularly used in classification. In this chapter, we introduce the notion of linear discriminant function and some of the important properties associated with it.

Keywords Linear classifier · Decision boundary · Linear separability · Nonlinear discriminant function · Linear discriminant function · Support vector machine · Perceptron

2.1 Introduction

We have seen in *Introduction* that a linear discriminant function $g(X)$ can be used as a classifier. The specific steps involved are as follows:

1. Consider a functional form for $g(X)$.
2. Using the two-class training data, learn $g(X)$. By learning $g(X)$ we mean obtaining the values of the coefficients of terms in $g(X)$.
3. Given a *test pattern* X_{test}, compute $g(X_{test})$. Assign X_{test} to C_- if $g(X_{test}) < 0$ else (if $g(X_{test}) > 0$) assign it to C_+.

2.1.1 Associated Terms [1–3]

We explain the associated concepts next

- **Training Dataset**:

   The training dataset or training set, $\mathcal{X}_{train}$, is a finite set given by

   $$\mathcal{X}_{train} = \{(X_1, C_1), (X_2, C_2), \cdots, (X_n, C_n)\}$$

M.N. Murty and R. Raghava, *Support Vector Machines and Perceptrons*,
SpringerBriefs in Computer Science, DOI 10.1007/978-3-319-41063-0_2

where X_i is the ith pattern (representation) given by $X_i = (x_{i1}, x_{i2}, \ldots, x_{il})$ for some finite l.

Even though it is possible to have more than two classes, we consider only two-class (*binary*) classification problems in this chapter. We will examine how to build a multiclass classifier based on a combination of binary classifiers later. So, Associated with pattern X_i is its class label C_i where $C_i \in \{C_-, C_+\}$.

- **Test Pattern**:

  A test pattern, X_{test} or simply X is an l-dimensional pattern which is not yet labeled.

- **Classifier**:

  A classifier *assigns a class label* to a *test/unlabeled pattern*.

We illustrate these notions with the help of a two-dimensional dataset shown in Fig. 2.1. We depict in the figure, a set of children and a set of adults. Each *child* is depicted using C and each *adult* using A. In addition there are four test patterns X_1, X_2, X_3, *and* X_4. Each pattern is represented by its *Height* and *Weight*.

In Fig. 2.1 three classifiers are shown, a *decision tree classifier*, an *LDF based classifier*, and a *nonlinear discriminant based classifier*.

Each of the three classifiers in the figure belongs to a different category. Here,

- The *Linear discriminant/classifier* depicted by the thin broken line is a *linear classifier*. Any point X falling on the left side of the line (or $g(X) < 0$) is a *child* and

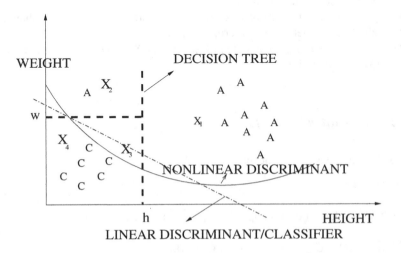

Fig. 2.1 An example dataset

a point X to the right (or $g(X) > 0$) is classified as *adult*.

- The *Nonlinear discriminant* shown by the curved line in the figure corresponds to a *nonlinear classifier*. An X such that $g(X) < 0$ is assigned the label *child*. If $g(X) > 0$, then X is assigned *adult*.

- The *decision tree* classifier depicted by the piecewise linear region in the figure is not linear and it could be called a *piecewise linear classifier*. It may be described by

  Adult : $(HEIGHT > h) \vee [(HEIGHT < h) \wedge (WEIGHT > w)]$.

  Child : $(HEIGHT < h) \wedge (WEIGHT < w)$.

In this simple case, test patterns X_1 *and* X_2 are assigned to class *Adult* or equivalently X_1 *and* X_2 are assigned the class label *Adult* by all the three classifiers.

Similarly, test pattern X_4 is assigned the label *child* by all the three classifiers. However, X_3 is assigned the label *adult* by the nonlinear discriminant-based classifier and the other two classifiers assign X_3 to class *child*.,

It is possible to extend these ideas to more than two-*dimensional* spaces. In high-dimensional spaces,

- the linear discriminant is characterized by a *hyperplane* instead of a line as in the two-dimensional case.

- the nonlinear discriminant is characterized by a *manifold* instead of a curve.

- the piecewise linear discriminant characterizing the decision tree classifier continues to be piecewise linear discriminant, perhaps involving a larger size conjunction. So, learning a decision tree classifier in high-dimensional spaces could be computationally prohibitive.

However, it is possible to classify X based on the value of $g(X)$ irrespective of the dimensionality of X (or the value of l). This needs obtaining an appropriate $g(X)$. In this chapter, we will concentrate on *linear classifiers*.

2.2 Linear Classifier [2–4]

A *linear classifier* is characterized by a *linear discriminant function* $g(X) = W^t X + b$, *where* $W = (w_1, w_2, \ldots, w_l)^t$ *and* $X = (x_1, x_2, \ldots, x_l)^t$. We assume without loss of generality that W *and* $X \in \mathbb{R}^l$ *and* $b \in \mathbb{R}$.

Note that both the components of W and X are in linear form in $g(X)$. It is also possible to express $g(X)$ as

$$g(X) = b + \sum_{i=1}^{l} w_i x_i$$

If we augment X and W appropriately and convert them into $l + 1$ dimensional vectors, we can have a more acceptable and simpler form for $g(X)$. The *augmented* form is given by $X_a = (1, x_1, \ldots, x_l)^t$ and $W_a = (b, w_1, \ldots, w_l)^t$, where X_a and W_a are augmented versions of X and W, respectively. Note that both X_a and W_a are $l + 1$ dimensional vectors.

Now

$$g(X) = W_a^t X_a = b + \sum_{i=1}^{l} w_i x_i$$

If we use the augmented vectors, then $g(X)$ satisfies the two properties of *linear systems* as shown below.

• **Homogeneity**: For $c \in \mathbb{R}$, $g(cX) = cg(X)$

$g(cX) = W_a^t(cX_a) = cW_a^t X_a = cg(X)$

• **Additivity**: For X_1 *and* $X_2 \in \mathbb{R}^l$, $g(X_1 + X_2) = g(X_1) + g(X_2)$

$g(X_1 + X_2) = W_a^t(X_{1a} + X_{2a}) = W_a^t X_{1a} + W_a^t X_{2a} = g(X_1) + g(X_2)$.

Note that if W and X are used in their l-dimensional form, then *homogeneity* and *additivity* are not satisfied. However, *convexity* is satisfied as shown below.

• **Convexity**: For some $\alpha \in [0, 1]$, $g(\alpha X_1 + (1 - \alpha)X_2) \le \alpha g(X_1) + (1 - \alpha)g(X_2)$

$g(\alpha X_1 + (1 - \alpha)X_2) = b + W^t(\alpha X_1 + (1 - \alpha)X_2)$

$= \alpha b + (1 - \alpha)b + \alpha W^t X_1 + (1 - \alpha)W^t X_2$

$= \alpha(b + W^t X_1) + (1 - \alpha)(b + W^t X_2) = \alpha g(X_1) + (1 - \alpha)g(X_2)$

• **Classification of augmented Vectors using** W_a:

We will illustrate classification of patterns using the augmented representations of the six patterns shown in Fig. 1.3. We show the augmented patterns in Table 2.1 along with these value of $W_a^t X_a$ for $W_a = (-14, 1, 5)^t$.

Table 2.1 Classification of augmented patterns using $W_a = (-14, 1, 5)^t$

Pattern number	Class label	1	x_1	x_2	$W_a^t X_a$
1	−	1	1	1	−8
2	−	1	2	2	−2
3	+	1	2	3	3
4	+	1	6	2	2
5	+	1	7	2	3
6	+	1	7	3	8

2.3 Linear Discriminant Function [2]

We have seen earlier in this chapter that a linear discriminant function is of the form $g(X) = W^t X + b$ where W is a column vector of size l and b is a scalar. $g(X)$ divides the space of vectors into three parts. They are

2.3.1 Decision Boundary

In the case of linear discriminant functions, $g(x) = W^t X + b = 0$ characterizes the *hyperplane (line in a two-dimensional case)* or the *decision boundary*. The decision boundary corresponding to $g(X)$ (DB_g) could also be viewed as

$$DB_g = \{X | g(X) = 0\}$$

2.3.2 Negative Half Space

This may be viewed as the set of all patterns that belong to C_-. Equivalently, the negative half space corresponding to $g(X)$ (NHS_g) is the set

$$NHS_g = \{X | g(X) < 0\} = C_-$$

2.3.3 Positive Half Space

This is the set of all patterns belonging to C_+. Equivalently, the positive half space corresponding to $g(X)$ (PHS_g) is given by

$$PHS_g = \{X | g(X) > 0\} = C_+$$

Note that each of these parts is a potentially infinite set. However, the training dataset and the collection of test patterns that one encounters are finite.

Fig. 2.2 Linearly separable
dataset

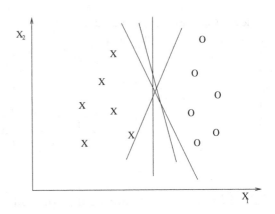

2.3.4 Linear Separability

Let $\mathcal{X}$ be a set of labeled patterns given by

$$\mathcal{X} = \{(X_1, C_1), (X_2, C_2), \dots, (X_n, C_n)\}.$$

We say the set $\mathcal{X}$ is *linearly separable* if there is a W and b such that $W^t X_i + b > 0$
if $C_i = C_+$ and $W^t X_i + b < 0$ if $C_i = C_-$ for $i = 1, 2, \dots, n$.

We can think of employing linear classifiers when the samples/set of patterns is
linearly separable. Consider the two-dimensional patterns shown in Fig. 2.2. They
are linearly separable. If they are linearly separable, then we can have infinite number
of LDFs associated as shown in the figure.

2.3.5 Linear Classification Based on a Linear Discriminant Function

A linear classifier is *abstracted* by the corresponding *ldf*, $g(X) = W^t X + b$. The
three regions associated with $g(X)$ are important in appreciating the classifier as
shown in Fig. 2.3.

1. **The decision boundary** or the hyperplane associated with $g(X)$ is the *separator*
   between the two classes, the *negative* and *positive* classes. Any point X on the
   decision boundary satisfies $g(X) = 0$.

   If X_1 and X_2 are two different points on the decision boundary, then

   $$W^t X_1 + b = W^t X_2 + b = 0 \implies W^t (X_1 - X_2) = 0.$$

   This means W **is orthogonal** to $(X_1 - X_2)$ or the line joining the two points X_1
   and X_2 or the decision boundary. So, W *is orthogonal to the Decision boundary*.

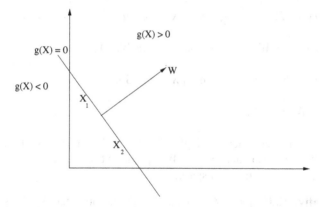

Fig. 2.3 Three regions associated with $g(X) = W^t X + b$

This means that there is a natural association between W and the decision boundary; in a sense if we specify one, the other gets fixed.

2. **The Positive Half Space**: Any pattern X in this region satisfies the property that $g(X) = W^t X + b > 0$. We can interpret it further as follows:

a. **Role of b**: We can appreciate the role of b by considering the value of $g(X)$ at the *origin*. Let $b > 0$ and X *is the origin*. Then $g(0) = W^t 0 + b = 0 + b = b > 0$. So, at the origin 0, $g(0) > 0$; hence the origin 0 is in the positive half space or PHS_g.

If $b > 0$, then the origin is in the positive half space of $g(X)$.

Now consider the situation where $b = 0$. So, $g(X) = W^t X + b = W^t X$. If X is at the origin, then $g(X) = g(0) = W^t 0 = 0$. So, the origin satisfies the property that $g(X) = 0$ and hence it is on the decision boundary.

So, if $b = 0$, then the origin is on the decision boundary.

b. **Direction of W**: Consider an LDF $g(X)$ where $b = 0$. Then $g(X) = W^t X$. If X is in the positive half space, then $g(X) = W^t X > 0$. We have already seen that W is orthogonal to the decision boundary $g(X) = 0$. Now we will examine whether W is oriented toward the positive half space or the negative half space.

If $b = 0$ and X is in the positive half space, then $g(X) = W^t X > 0$. Now relate $W^t X$ with the cosine of the angle between W and X. We have

$cosine(W, X) = \frac{W^t X}{||W|| \, ||X||} \Rightarrow W^t X = cosine(W, X) \, || \, W \, || \, || \, X \, ||.$

So, given that $W^t X > 0$, we have $cosine(W, X) \, || \, W \, || \, || \, X \, || > 0$

We know that $|| \, W \, || > 0$ and $|| \, X \, || > 0$. So,

$cosine(W, X) > 0.$

This can happen when the angle, θ, between W and X is such that $-90 < \theta < 90$ which can happen when W is pointing toward the positive half space as X is in the positive half space.

3. **The Negative Half Space**: Any point X in the negative half space is such that $g(X) < 0$. Again if we let $b = 0$ and consider a pattern, X, in the negative class, then $W^t X < 0$. This means the angle, θ, between X and W is such that $90 < \theta < -90$. This also ratifies that W points toward the positive half space.

Further, note that for $b < 0$ and X in the negative half space, $g(X) = W^t X + b < 0$ and evaluated at the origin, $g(0) = W^t 0 + b = b < 0$. So, *if $b < 0$, then the origin is in the negative half space.*

So, the roles of W and b in the LDF $g(X) = W^t X + b$ are given by

- *The value of b decides the location of the origin.* The origin is in the PHS_g if $b > 0$; it is in the NHS_g if $b < 0$ and the origin is on the decision boundary if $b = 0$. It is illustrated in Fig. 2.4

Note that there are patterns from two classes and the samples are linearly separable. There are three linear discriminant functions with different b values and correspondingly the origin is in the negative space in one case ($x_1 = x_2 - C_1$), on the decision boundary in the second case ($x_1 = x_2$) and it is in the positive space

Fig. 2.4 Three decision boundaries with same W

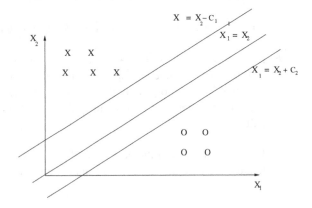

in the third ($x_1 = x_2 + C_2$). However, W is the same for all the three functions as the decision boundaries are all parallel to each other.

- *W is orthogonal to the decision boundary and it points toward the positive half space of g* as shown in Fig. 2.3.

2.4 Example Linear Classifiers [2]

It is possible to show that the *MDC, Naïve* Bayes classifier and others are linear classifiers. Consider

2.4.1 Minimum-Distance Classifier (MDC)

In the case of MDC we assign X to C_- if

$$|| X - m_- ||^2 < || X - m_+ ||^2 \Rightarrow$$
$$X^t X + m_-^t m_- - 2m_-^t X < X^t X + m_+^t m_+ - 2m_+^t X.$$

We can simplify by canceling $X^t X$ that is common to both sides and bringing all the terms to the left-hand side, we get

assign X to C_- if $(m_+ - m_-)^t X + \frac{1}{2}(m_-^t m_- - m_+^t m_+) < 0$.

This is the same as assigning X to C_- if $W^t X + b < 0$ where

$W = (m_+ - m_-)$ and $b = \frac{1}{2}(m_-^t m_- - m_+^t m_+)$.

So, MDC is a linear classifier characterized by an LDF of the form $W^t X + b$.

2.4.2 Naïve Bayes Classifier (NBC)

In the case of NBC, we have

$$P(C_-|X) = \prod_{i=1}^{l} P(x_i|C_-)P(C_-)$$

and

$$P(C_+|X) = \prod_{i=1}^{l} P(x_i|C_+)P(C_+)$$

We assign X to C_- if $P(C_-|X) > P(C_+|X)$ or equivalently when

$$\prod_{i=1}^{l} P(x_i|C_-)P(C_-) > \prod_{i=1}^{l} P(x_i|C_+)P(C_+).$$

By applying logarithm both sides and rearranging terms, we have

$$\sum_{i=1}^{l} n_i log \frac{P(x_i|C_-)}{P(x_i|c_+)} + log \frac{P(C_-)}{P(C_+)} > 0$$

where n_i is the number of times the feature x_i occurred in X. If X is a binary pattern, then n_i is either 1 or 0. If X is a document, then n_i is the number of times term x_i occurred in X.

So, we assign X to C_- if

$$\sum_{i=1}^{l} w_i n_i + b > 0$$

where

$w_i = log \frac{P(x_i|C_-)}{P(x_i|c_+)}$, $b = log \frac{P(C_-)}{P(C_+)}$.

So, *Naïve Bayes Classifier* is a linear classifier.

2.4.3 Nonlinear Discriminant Function

It is possible to view a nonlinear discriminant function as a linear discriminant function in a higher dimensional space. For example, consider the two-dimensional dataset of six patterns shown in Fig. 1.6.

We have seen that a nonlinear discriminant function given by $x_1^2 + 32x_2 - 76$ can be used to classify the six patterns.

Here, X is a two-dimensional column vector given by $X = (x_1, x_2)^t$. However, if we map it to a six-dimensional representation given by $\phi(X) = (1, x_1, x_2, x_1^2, x_2^2, x_1x_2)^t$ where

$\phi_1(X) = 1$, $\phi_2(X) = x_1$, $\phi_3(X) = x_2$, $\phi_4(X) = x_1^2$, $\phi_5(X) = x_2^2$, $\phi(X) = x_1x_2$.

So, ϕ is a mapping from $\mathbb{R}^2$ to $\mathbb{R}^6$ such that

$\phi : (x_1, x_2)^t \rightarrow (1, x_1, x_2, x_1^2, x_2^2, x_1x_2)^t$.

Then the nonlinear discriminant function $x_1^2 + 32x_2 - 76$ in $\mathbb{R}^2$ is linear in $\mathbb{R}^6$ corresponding to the $\phi(X)$ space.

If we choose $W = (-76, 0, 32, 1, 0, 0)$ then, $g(X) = W^t\phi(X)$ which is a linear discriminant function in $\phi(X)$.

References

1. Bishop, C.M.: Pattern Recognition and Machine Learning. Springer (2006)
2. Duda, R.O., Hart, P.E.: Pattern Classification and Scene Analysis. Wiley (1970)
3. Fukunaga, K.: Introduction to Statistical Pattern Recognition. Academic Press (2013)
4. Zhao, W., Chellappa, R., Nandhakumar, N.: Empirical performance analysis of linear discriminant classifiers, In: Proceedings of Computer Vision and Pattern Recognition, 25–28 June 1998, pp. 164–169. Santa Barbara, CA, USA (1998)

Chapter 3
Perceptron

Abstract Perceptron is a well-known classifier based on a linear discriminant function. It is intrinsically a binary classifier. It has been studied extensively in its early years and it provides an excellent platform to appreciate classification based on Support Vector Machines. In addition, it is gaining popularity again because of its simplicity. In this chapter, we introduce perceptron-based classification and some of the essential properties in the context of classification.

Keywords Perceptron · Learning algorithm · Optimization · Classification · Order of perceptron · Incremental computation

3.1 Introduction

Perceptron [1–3] is a well-known and is the first binary classifier based on the notion of linear discriminant function. The *perceptron learning algorithm* learns a linear discriminant function $g(X) = W^t X + b$ from the training data drawn from two classes. Specifically, it learns W and b. In order to introduce the learning algorithm, it is convenient to consider the augmented vectors which we have seen in the previous chapter.

Recall the augmented pattern X_a of the pattern X given by
$X_a = (1, x_1, x_2, \ldots, x_l)$ and the corresponding weight vector W_a
$W_a = (b, w_1, w_2, \ldots, w_l)$.
We know that $g(X) = W_a^t X_a$ and we assign X to class C_- if $g(X) < 0$ and assign X to C_+ if $g(X) > 0$.

We assume that there is no X such that $g(X) = 0$ or equivalently there is no X on the decision boundary. This assumption also means that the classes are *linearly separable*.

It is convenient to consider yX where y is the class label of pattern X. Further, we assume that
$y = -1\ if\ X \in C_-$ and
$y = +1\ if\ X \in C_+$.

M.N. Murty and R. Raghava, *Support Vector Machines and Perceptrons*,
SpringerBriefs in Computer Science, DOI 10.1007/978-3-319-41063-0_3

Table 3.1 Classification based on $g(yX)$ using $W_a = (-14, 1, 5)^t$

Pattern number	Class label	1	x_1	x_2	$W_a^t y X_a$
1	−	−1	−1	−1	8
2	−	−1	−2	−2	2
3	+	1	2	3	3
4	+	1	6	2	2
5	+	1	7	2	3
6	+	1	7	3	8

We have

if $X \in C_-$, then $g(X) = W_a^t X_a < 0$. Hence $g(yX) = W_a^t y X_a > 0$ as $y = -1$.
If $X \in C_+$, then $g(X) = W_a^t X_a > 0$. So, $g(yX) = W_a^t y X_a > 0$ as $y = +1$.

So, $g(yX) > 0$ irrespective of whether $X \in C_-$ or $X \in C_+$. This view simplifies the learning algorithm. Now we consider augmented patterns shown in Table 2.1 and show them in Table 3.1 using yX_a for each X_a.

Note that the vector $(-14, 1, 5)^t$ classifies all the yX_as correctly.

In the rest of this chapter we use the following notation, for the sake of brevity and simplicity.

- We use W for W_a with the assumption that b is the first element in W
- We use X for yX_a assuming that X is augmented by adding 1 as the first component and the vector X_a is multiplied by y; we call the resulting vector X.
- We *learn* W from the training data.
- We use *Perceptron learning algorithm* for learning W.

We discuss the algorithm and its analysis next.

3.2   Perceptron Learning Algorithm [1]

1. Initialize i to be 0 and W_i to be the *null vector*, **0**.
2. For $k = 1$ *to* n do
   if W_i misclassifies X_k, that is if $W_i^t X_k \leq 0$, then $W_{i+1} = W_i + X_k$; set $i = i + 1$.
3. Repeat Step 2 till the value of i does not change over an entire iteration (or epoch) over all the n patterns.

3.2.1   Learning Boolean Functions

We can illustrate the algorithm with the help of a boolean function; we consider the boolean *or* function. The *truth table* is shown in Table 3.2.

Table 3.2 Truth table of inclusive OR

x_1	x_2	$x_1 \vee x_2$
0	0	0
0	1	1
1	0	1
1	1	1

Table 3.3 Classification based on vectors yX_a

Pattern number	Class label 1		x_1	x_2
1	−1	−1	0	0
2	1	1	0	1
3	1	1	1	0
4	1	1	1	1

We view it as a two-class problem where the output 0 is taken as indicating the negative class and output 1 is an indicator of the positive class. After augmenting and multiplying with the class label of $y = -1$ or $+1$, respectively, for the negative or positive class patterns, we have the data shown in Table 3.3.

We start with $W_0 = (0, 0, 0)^t$. The stepwise updates on W are:

1. W_0 misclassifies the first vector $(-1, 0, 0)^t$ as the dot product between them is 0. So, $W_1 = W_0 + (-1, 0, 0)^t = (-1, 0, 0)^t$.
2. W_1 misclassifies the second pattern $(1, 0, 1)^t$ as the dot product is $-1 < 0$. So, $W_2 = W_1 + (1, 0, 1)^t = (0, 0, 1)^t$.
3. W_2 misclassifies the third pattern $(1, 1, 0)^t$; the dot product is 0. So, $W_3 = W_2 + (1, 1, 0)^t = (1, 1, 1)^t$.
4. Note that W_3 classifies the fourth pattern $(1, 1, 1)^t$ correctly; the dot product is $3 > 0$. Now we go through the patterns again starting from the first pattern. The weight W_3 fails to classify the first pattern $(-1, 0, 0)^t$ as the dot product is -1. So, $W_4 = W_3 + (-1, 0, 0)^t = (0, 1, 1)^t$.
5. Note that W_4 fails to classify the first pattern correctly even though it classifies patterns numbered 2, 3, and 4. So, $W_5 = W_4 + (-1, 0, 0)^t = (-1, 1, 1)^t$.
6. W_5 misclassifies the second pattern and so $W_6 = W_5 + (1, 0, 1)^t = (0, 1, 2)^t$.
7. W_6 misclassifies the first pattern after having correctly classified patterns numbered 3 and 4, so $W_7 = W_6 + (-1, 0, 0)^t = (-1, 1, 2)^t$.
8. W_7 misclassifies the third pattern; hence $W_8 = W_7 + (1, 1, 0)^t = (0, 2, 2)^t$.
9. W_8 classifies the first pattern incorrectly. So, $W_9 = W_8 + (-1, 0, 0)^t = (-1, 2, 2)^t$. Note that W_9 classifies all the four patterns correctly. So, the discriminant function $g(X)$ is of the form $g(X) = (-1, 2, 2)(1, x_1, x_2)^t$. Hence the decision boundary is of the form $2x_1 + 2x_2 = 1$.

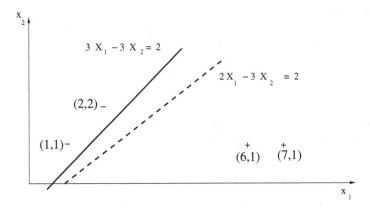

Fig. 3.1   Order dependence of the algorithm

3.2.2   W Is Not Unique

The W vector obtained may depend on the order in which we process the data points. As soon as we get a W that classifies all the patterns correctly, we stop the iterations. Consider the dataset shown in Fig. 3.1.

There are four patterns. They are from two classes as shown below:
Negative class: $(1, 1)^t$, $(2, 2)^t$
Positive Class: $(6, 1)^t$, $(7, 1)^t$

The augmented patterns after multiplying with y are:
Negative class: $X_1 = (-1, -1, -1)^t$, $X_2 = (-1, -2, -2)^t$
Positive class: $X_3 = (1, 6, 1)^t$, $X_4 = (1, 7, 1)^t$

- If we use the patterns in the given order, that is X_1, X_2, X_3, *and* X_4, then $W_0 = (0, 0, 0)^t$ and the algorithm stops at $W_4 = (-2, 2, -3)^t$. So, the decision boundary is $2x_1 - 3x_2 = 2$ and is depicted using a broken line.
- If we use the order, X_4, X_3, X_1, *and* X_2, then we get $W_4 = (-2, 3, -3)^t$ starting with $W_0 = (0, 0, 0)^t$ and W_4 correctly classifies all the four patterns; here the decision boundary is $3x_1 - 3x_2 = 2$.

3.2.3   Why Should the Learning Algorithm Work?

- *Algebraic Argument:*
  If W_i has misclassified X_k, then $W_i^t X_k \leq 0$. Note that
  $W_{i+1}^t X_k = (W_i + X_k)^t X_k = W_i^t X_k + X_k^t X_k$

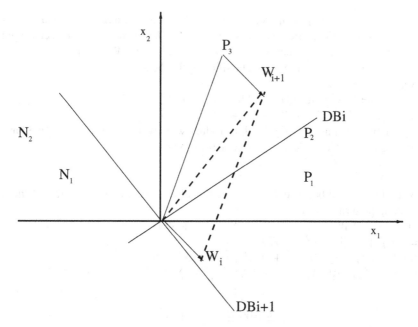

Fig. 3.2 Geometric support for the update of W_i

So, $W_{i+1}^t X_k > W_i^t X_k$ because $X_k^t X_k = \| X_k \|^2 > 0$ as squared euclidean norm is positive.

As a consequence W_{i+1} *is better suited to classify* X_k *than* W_i and $W_{i+1}^t X_k$ can be positive even if $W_i X_k$ is not.

- *Geometric Argument:*

  We illustrate how W_i is updated when a pattern is misclassified by it. Consider the two-dimensional dataset shown in Fig. 3.2.

  We have W_i and the corresponding decision boundary DBi as shown in the figure. Clearly W_i misclassifies P_3 and W_{i+1} is obtained by adding P_3 to W_i by parallelogram completion as shown in the figure.

  Note that W_{i+1} (indicated by the broken line) correctly classifies P_3. Also note the corresponding decision boundary DB_{i+1} that is orthogonal to W_{i+1}.

3.2.4 Convergence of the Algorithm

It is possible to show that the perceptron learning algorithm *converges to a correct weight vector in a finite number of iterations* if the classes are linearly separable. The proof goes as follows:

- Let $\{X_1, X_2, \ldots, X_n\}$ be the set of patterns obtained after augmentation and multiplication with the respective class label, -1 or +1. Let W be the correct weight vector with unit norm. That is, $W^t X_i > 0$, for all i and $\| W \| = 1$.
- $W_1 = 0$. Let the first pattern misclassified by W_1 be X^1.
  So, $W_2 = W_1 + X^1$.
- If for some k, X^k is misclassified by W_k, then $W_{k+1} = W_k + X^k$.
- So, $W_{k+1} = W_1 + X^1 + X^2 + \cdots + X^k$; $W_1 = 0$.
  By ignoring W_0 and taking the dot product with W on both the sides, we have
  $W^t W_{k+1} = W^t (X^1 + X^2 + \cdots + X^k) > k\alpha$ where
  $$\alpha = \frac{min}{X_i} \; W^t X_i$$
  Note that $\alpha > 0$ because $W^t X_i > 0$, for all X_i because W classifies all the n training patterns correctly.
- Observe that $W_{k+1}^t W_{k+1} = (W_k + X^k)^t (W_k + X^k)$
  $= \| W_k \|^2 + \| X^k \|^2 + 2 W_k^t X^k$
- So, $W_{k+1}^t W_{k+1} \leq \| W_k \|^2 + \| X^k \|^2$ as $W_k^t X^k \leq 0$.
- $\Rightarrow W_{k+1}^t W_{k+1} \leq \| W_{k-1} \|^2 + \| X^{k-1} \|^2 + \| X^k \|^2$
-

$$\leq \| W_1 \|^2 + \sum_{i=1}^{k} \| X^i \|^2$$

- Noting that $W_1 = 0$, we have
  $$W_{k+1}^t W_{k+1} = \| W_{k+1} \|^2 < k\beta \text{ where } \beta = \frac{max}{X_i} \| X^i \|^2$$
- Let θ be the angle between W_{k+1} and W (note that $\| W \| = 1$). So, we have
- $1 \geq cos\theta = \frac{W^t W_{k+1}}{\|W\| \|W_{k+1}\|} = \frac{W^t W_{k+1}}{\|W_{k+1}\|} > \frac{k\alpha}{\sqrt{k\beta}} \Rightarrow \frac{k\alpha}{\sqrt{k\beta}} < 1 \Rightarrow k < \frac{\beta}{\alpha^2}$
- Note that both α and β are finite and positive. So, k is finite and the algorithm has finite and deterministic convergence.
- However, $\alpha \to 0 \Rightarrow k \to \infty$ and $\alpha \to 0$ when one of the patterns is close to being orthogonal to W.
- So, **perceptron learning algorithm converges to a correct W within a finite number of iterations, k, over the data if the classes are linearly separable**. However, the value of k may increase if a pattern is very closely located to the decision boundary or more appropriately it is very close to being orthogonal to W.

3.3 Perceptron Optimization

We can view perceptron as *minimizing a cost function $J(W)$* associated with the patterns of the form $y X_a$ and we call them X as mentioned earlier in the current chapter.

- Specifically, $J(W)$ may be specified based on the patterns Xs, misclassified by W. It is

$$J(W) = - \sum_{j:W^tX_j \leq 0} W^t X_j$$

- Note that $J(W)$ is nonnegative as it is based on the patterns misclassified by W; every X_j misclassified by W is such that $W^t X_j \leq 0$.
- If a W classifies all the X_js correctly, then $J(W) = 0$.
- So, learning optimal W corresponds to obtaining the right W that minimizes $J(W)$.
- A simple approach based on the well-known *gradient descent approach* will be adequate here.

$$\triangledown J(W) = - \sum_{j:W^tX_j \leq 0} X_j$$

- So, W_{k+1} is obtained from W_k using the gradient descent approach by subtracting the negative of the gradient from W_k. That is
-

$$W_{k+1} = W_k - \eta \triangledown J(W_k) = W_k + \eta \sum_{j:W^tX_j \leq 0} X_j$$

- This is the so-called *batch update rule* as it adds to W_k all the patterns X_js that are misclassified by W_k to get W_{k+1}.

3.3.1 Incremental Rule

A variant of the batch update rule is obtained by letting $\eta = 1$ and consider the first pattern, X_j that is misclassified by W_k. Let us call such an X_j by X^k.

So, the incremental rule with a fixed step size ($\eta = 1$) is given by $W_{k+1} = W_k + X^k$ where X^k is the first pattern misclassified by W_k. Note that this is *precisely the step we followed in the perceptron learning algorithm* for which we have seen the convergence.

3.3.2 Nonlinearly Separable Case

An important observation is that the *perceptron learning algorithm* can be used to learn a nonlinear discriminant provided the form of the nonlinear function is known.

- For example, consider the one-dimensional two-class dataset shown in Table 3.4
- Let the form of the discriminant function be $a + bx + cx^2$. This can be viewed as a function of the form $W^t \phi(X)$ where $W = (a, b, c)^t$ and $\phi(x) = (x, x^2)$; so $\phi_a(x) = (1, x, x^2)$. Using the form of $\phi_a(x)$ and class label y, the data can be shown as in Table 3.5 after augmenting and multiplying with the class label.

Table 3.4 Linear
discriminant function

Pattern number	x	Class label
1	1	+
2	−1	+
3	2	+
4	−2	+
5	3	−
6	4	−
7	−3	−
8	−4	−

Table 3.5 Data augmented
and multiplied by the class
label value

Pattern number	1	$\phi_1(x) = x$	$\phi_2(x) = x^2$
1	1	1	1
2	1	−1	1
3	1	2	4
4	1	−2	4
5	−1	−3	−9
6	−1	−4	−16
7	−1	3	−9
8	−1	4	−16

- We start with $W_0 = (0, 0, 0)^t$ and go through the perceptron learning algorithm. We get the vector $(11, 0, -2)^t$ that classifies all the eight patterns correctly ($W^t X > 0$). The decision boundary is given by $11 - 2x^2 = 0$ or equivalently, $x^2 = \frac{11}{2}$.
- So, *perceptron learning algorithm* can be used to find the weight vector W even when the original problem is nonlinear. We need to transform it into a high-dimensional space (ϕ space) appropriately.
- In the above example, the one-dimensional value x is mapped to a two-dimensional vector in the ϕ space, where $\phi_1(x) = x$ *and* $\phi_2(x) = x^2$.

3.4 Classification Based on Perceptrons [2]

Perceptron can be used as a binary classifier based on learning the weight vector W. It is such that

- If $g(X)$ is linear in X, then we can use the form $g(X) = W^t X$. We can *augment* W and X appropriately to take care of b in $W^t X + b$ *implicitly*.
- If $g(X)$ is nonlinear in X, then we can use $g(X) = W^t \phi(X)$ by transforming the discriminant function that is nonlinear in X into a function that is linear in $\phi(X)$.

There are a couple of important properties of the W vector obtained using the perceptron learning algorithm.

1. **Weight Vector and the Training Patterns**: We have observed earlier that

$$W_{k+1} = X^1 + X^2 + \cdots + X^k = \sum_{i=1}^{k} X^i$$

This means that the final weight vector may be viewed as a sum of the training patterns that were misclassified at various steps.

2. **Weight Vector and Feature Selection**: It is easy to observe that value of w_i indicates the importance of the ith feature.

For example, consider the vector $W = (0.01, -34, 0, 17.5)^t$. It indicates that the second and the fourth features are important in that order and and the third feature can be ignored. Also the first feature can almost be ignored.

These properties of W are shared by all the classifiers based on linear discriminants including SVMs and perceptrons.

3.4.1 Order of the Perceptron

An important characterization of perceptron is its *order*. Intuitively, it signifies the inherent computational need. We can explain it in detail as follows:

- **Exclusive OR (xor)**
  It is well known that $xor(x_1, x_2)$ is not linear in x_1 *and* x_2. It is possible to have a linear representation in a high-dimensional space involving $\phi_1(X) = x_1$, $\phi_2(X) = x_2$, *and* $\phi_3(X) = x_1x_2$
  Specifically, $xor(x_1, x_2) = x_1 + x_2 - 2x_1x_2$. The truth table is shown in Table 3.6.

- **Odd Parity Predicate of size 3**
  Consider the predicate $x_1 + x_2 + x_3$ *is odd*. It can be represented as a linear discriminant function in the ϕ space where $\phi(X)$ is given by:
  $\phi(X) = (x_1, x_2, x_3, x_1x_2, x_2x_3, x_1x_3, x_1x_2x_3)^t$. Let $g(X) = x_1 + x_2 + x_3 - 2x_1x_2 - 2x_2x_3 - 2x_1x_3 + 4x_1x_2x_3$. The truth table of the *odd parity predicate* is given in Table 3.7.

Table 3.6 Truth table of Exclusive OR

x_1	x_2	$xor(x_1, x_2)$	$x_1 + x_2 - 2x_1x_2$
0	0	0	0
0	1	1	1
1	0	1	1
1	1	0	0

Table 3.7 Truth table of the odd parity predicate

x_1	x_2	x_3	*odd parity*(X)	$g(X)$
0	0	0	0	0
0	0	1	1	1
0	1	0	1	1
0	1	1	0	0
1	0	0	1	1
1	0	1	0	0
1	0	1	0	0
1	1	1	1	1

- The common features of *xor* and *odd parity* predicates are:

  1. Both are *nonlinear* in the input features; *xor* is nonlinear in x_1 *and* x_2; and *odd parity* is nonlinear in $x_1, x_2,$ *and* x_3.
  2. In both the cases the corresponding linear representations are in higher dimensional spaces involving the *minterms* of the respective input features.
  3. Note that *xor*(x_1, x_2) involves $x_1, x_2,$ *and* x_1x_2, all the three minterms. Similarly, *odd parity*(x_1, x_2, x_3) employs $x_1, x_2, x_3, x_1x_2, x_2x_3, x_1x_3, x_1x_2x_3$, all the seven minterms.
  4. If there are l boolean input variables $x_1, x_2, \cdots, x_l$, then the total number of minterms is

$$\binom{l}{1} + \binom{l}{2} + \cdots + \binom{l}{l} = 2^l - 1$$

  5. Another property is that all the minterms of the same size have the same coefficient. For example, in *odd parity*(x_1, x_2, x_3), $x_1, x_2,$ *and* x_3 each has a coefficient 1, $x_1x_2, x_2x_3,$ *and* x_1x_3 have a coefficient -2 each, and $x_1x_2x_3$ has a coefficient 4.

- The *support* of a minterm is its size. For example, the support of x_1 or x_2 is 1; support of x_2x_3 is 2; and the support of $x_1x_2x_3$ is 3.
- We can use *support* to define the order of the predicate.
  The *order of a predicate* is the smallest number k for which we can find a set of features (ϕs), possibly in a higher dimensional space where every ϕ has a support less than or equal to k.
  For example, the order of *xor*(x_1, x_2) is 2 and that of *odd parity*(x_1, x_2, x_3) is 3.
  If a binary image has l pixels, with black pixel having value 1 and white pixel with value 0, then the predicate *the image has at least one black pixel* is $x_1 + x_2 + \cdots + x_l > 0$. So, the order of this predicate is 1. Note that this representation involves x_1 to x_l where each feature has a support of 1.

3.4.2 Permutation Invariance

Some predicates are invariant to permutations. For example, $xor(x_1, x_2)$ is the same
as $xor(x_2, x_1)$. Similarly *odd parity*(x_1, x_2, x_3) and the *image having at least one
black pixel* are all *permutation invariant*.
However, *a black pixel occurs before any white pixel* is not permutation invariant.

- **Positive Normal Form**:
  It is possible to represent any boolean function in terms of minterms only. Such a
  representation is called the *positive normal form*.
  For example, $xor(x_1, x_2) = x_1\bar{x}_2 + \bar{x}_1 x_2$. We can replace $\bar{x}$ by $1 - x$, to get
  $xor(x_1, x_2) = x_1(1 - x_2) + (1 - x_1)x_2 = x_1 + x_2 - 2x_1 x_2$
- **Coefficients of Minterms**:
  It is possible to show that a permutation invariant predicate can be represented
  in a linear form using only minterms, where minterms of the same size (or same
  support) have the same coefficient.
  We have seen this property earlier with respect to xor and *odd parity*. Similarly,
  the predicate *the image has at least one black pixel* has a simpler representation
  where only minterms of size 1 are used and the coefficient is 1 for all these
  minterms. So, there is no nonlinearity here.

3.4.3 Incremental Computation

Some of the boolean functions have simpler (or linear) and others have complex (or
nonlinear) representations. The associated properties are

- **Incremental**:
  If we consider an image of l pixels, then the predicate *the image has at least one
  black pixel* can be incrementally computed. The corresponding $g(X)$ is

$$g(X) = x_1 + x_2 + \cdots + x_l.$$

It is possible to represent a larger image with $l + p$ pixels using
$h_1(X) = x_1 + x_2 + \cdots + x_l$ and
$h_2(X) = x_{l+1} + x_{l+2} + \cdots + x_{l+p}$
$g(X) = h_1(X) + h_2(X).$

Such a formulation permits both incremental updation and also a suitable *divide-
and-conquer* approach for efficient computation. If required, $g(X)$ could be computed based on $h_1(X), h_2(X), \ldots,$ *and* $h_m(X)$ using
$g(X) = h_1(X) + h_2(X) + \cdots + h_m(X).$

- **Non-incremental**:
  However, predicates like *odd parity* and *xor* are nonlinear and cannot be computed incrementally as the order of such perceptrons increases with the increase in the size of X.
  Predicates like *odd parity* involve usage of all the minterms and so are highly nonlinear. So, it is very difficult to extend the computation incrementally and/or using a divide-and-conquer approach.

3.5   Experimental Results

We illustrate the working of Perceptron using the **Iris dataset**. This dataset has the following characteristics:

- There are **five features** associated with each of data points/flowers.
- The first feature is **Sepal Length** of the flower in centimeters.
- The second feature is **Sepal Width** of the flower in centimeters.
- The third feature is **Petal Length** of the flower in centimeters.
- The fourth feature is **Petal Width** of the flower in centimeters.
- The fifth feature associated with each flower is a dependent feature; it can assume one of three class labels from the Iris family of flowers. The class labels are **Setosa, Versicolour, and Virginica**.

So there are three classes and each class is represented by a collection of 50 flowers that are characterized by the five-dimensional vectors as specified before. It is a popularly used dataset and it is well known that Setosa and Versicolour are linearly separable in the four-dimensional feature space corresponding to the first four features. Similarly, the pair Setosa and Virginica are also linearly separable. However, Versicolour and Virginia are not linearly separable. In this chapter, we considered the linearly separable pair of Setosa and Versicolour for our experiments.

We have used 60 patterns out of 100 for training and the remaining 40 patterns for testing. Using the training data, we have learnt the augmented weight vector $(1, 1.2999, 4.1, -5.2, -2.1999)^t$. We have classified the test patterns using this weight vector and found that all of them are correctly classified; so, the accuracy of the perceptron classifier is 100 %.

- **Reduced Dimensional Space**
  In order to test the behavior of the perceptron classifier, we have conducted experiments using a subset of two features at a time. So, there are **six** possibilities. We use ij to represent that features i and j $(i \neq j)$ are used in the experiment. We provide the corresponding results in Table 3.8 based on the patterns from Setosa and Versicolour.

Table 3.8 Results of the perceptron classifier using two features at a time

Experiment number	Features	Weight vector	Testing accuracy (%)
1	1 and 2	$(2, -5.9, 9.2)^t$	97.5
2	1 and 3	$(2, 3.4, -9.1)^t$	100
3	1 and 4	$(2, 0.5, -5.6)^t$	100
4	2 and 3	$(1, 4.1, -5.2)^t$	100
5	2 and 4	$(0, 0.3, -1.2)^t$	100
6	3 and 4	$(2, -0.5, -0.8)^t$	100

Table 3.9 Results of the perceptron classifier on the handwritten digits

Part number	Rows	Testing accuracy (%)
1	1–4	Not linearly separable
2	5–8	99.85
3	9–12	99.55
4	13–16	99.55

- **Handwritten Digits**
  In this experiment, we have considered handwritten digits. A collection of 1000 handwritten 0s and 1000 handwritten 1s are collected. This data is divided into 667 for training and 333 for test in each class; the two classes are 0 and 1. Each pattern is a matrix of 16 rows and 12 columns. We divide each pattern, both training and test, into four parts. Each part consists of four rows ($4 \times 12 = 48$ pixels). The first part consists of rows 1–4; the second part has rows 5–8; rows 9–12 are in the third part; and the fourth part has rows 13–16. Results obtained are shown in Table 3.9.

3.6 Summary

1. Perceptron is an important classifier based on *linear discriminant functions*.
2. The linear discriminant function is of the form $g(X) = W^t X$ and W can be learnt using the *perceptron learning algorithm*.
3. If the function is nonlinear in the input variables, then we can map X to a high-dimensional space ($\phi(X)$).
4. A nonlinear function in X can be represented as a linear function in $\phi(X)$ such that $g(X) = W^t \phi(X)$. W can be learnt using the perceptron learning algorithm here also.
5. The perceptron learning algorithm can be used to get the correct W when the classes are linearly separable in X (input space) or in $\phi(X)$ (a high-dimensional) space.
6. Even subsets of features might be adequate to learn the classifier.

7. Any boolean function could be represented as a linear discriminant in the space of the minterms or in its *positive normal form*.
8. It is possible to show that the perceptron learning algorithm converges in a finite number of iterations to a correct W.
9. However, W obtained could be different for different runs with different processing orders.
10. It is possible to show that the perceptron criterion function is minimized to get a W that classifies all the patterns correctly.
11. The W vector is obtained in an iterative manner by adding the previous weight vector to a pattern misclassified by it. W is a sum of the training data points that are misclassified by earlier weight vectors.
12. The most important property of perceptrons is the *order of the perceptron*. It characterizes the computational difficulty associated with the perceptron.

References

1. Minsky, M.L., Papert, S.: Perceptrons: An Introduction to Computational Geometry, MIT Press (1988)
2. Nielsen, M.: Neural Networks and Deep Learning, Online Book (2016)
3. Chaudhuri, S., Tewari, A.: Perceptron like algorithms for online learning to rank, arXiv:1508.00842 (2015)

Chapter 4
Linear Support Vector Machines

Abstract Support vector machine (SVM) is the most popular classifier based on a linear discriminant function. It is ideally suited for binary classification. It has been studied extensively in several pattern recognition applications and in data mining. It has become a baseline standard for classification because of excellent software packages that have been developed systematically over the past three decades. In this chapter, we introduce SVM-based classification and some of the essential properties related to classification. Specifically we deal with linear SVM that is ideally suited to deal with linearly separable classes.

Keywords Linear SVM · Perceptron and SVM · Maximum margin · Dual problem · Binary classifier · Multiclass classification

4.1 Introduction

Support vector machine (SVM) [1–5] can be used as a binary classifier based on a *linear discriminant function*. In this sense it resembles the perceptron.

4.1.1 Similarity with Perceptron

1. Both perceptron and SVM can be seen as employing the linear discriminant function of the form $W^t X + b$.
2. In the case of perceptron, if the classes are *linearly separable* then it is possible to get more than one W as shown in Fig. 3.1. In theory, there could be *infinite solutions or W vectors*. In the case of SVM, we constrain the W to be a globally optimal solution of a well-formulated optimization problem. So, W is unique.
3. If there is no linear discriminant in the input space or in the given variables, then it is possible to get a linear discriminant in a high-dimensional space. We have seen that in the case of boolean functions, we can transform any function into a linear form in the space of all possible minterms.

© The Author(s) 2016
M.N. Murty and R. Raghava, *Support Vector Machines and Perceptrons*,
SpringerBriefs in Computer Science, DOI 10.1007/978-3-319-41063-0_4

4. For example, $xor(x_1, x_2)$ is not linear in (x_1, x_2). However, it is linear in (x_1, x_2, x_1x_2) as examined in the previous chapter.

5. Let $X = (x_1, x_2)^t$ be a two-dimensional vector and let $\phi : \mathbb{R}^2 \to \mathbb{R}^5$ given by $\phi(X) = (1, x_1, x_2, x_1^2, x_2^2, x_1x_2)$. Then $g(X) = a_0 + a_1x_1 + a_2x_2 + a_3x_1^2 + a_4x_2^2 + a_5x_1x_2$ is a *quadratic function* in $\mathbb{R}$, the *input space*, and $g(\phi(X)) = a_0 + a_1x_1 + a_2x_2 + a_3x_1^2 + a_4x_2^2 + a_5x_1x_2$ is a *linear function* in the 5-dimensional $\phi(X)$ space, called the *feature space*.

6. SVM and perceptron are *linear classifiers*.

7. Both SVM and perceptron are inherently binary classifiers. They can be extended to deal with multiclass classification using similar techniques which we will discuss later.

4.1.2   Differences Between Perceptron and SVM

1. **W Vector**:
   Perceptron can converge to different W vectors based on the order in which the training patterns are processed.
   However, SVM will produce the same W.

2. **Optimization**:
   Perceptron criterion function, $J(W)$. has value 0 if all the training patterns are classified correctly by W. In other words, $W^t X_i > 0$ for all i. So, multiple solutions, or W vectors could exist that lead to the same error.
   It is possible to show, on the contrary, that the SVM criterion function will result in the same W vector. Here, the W vector corresponds to the *decision boundary that maximizes separation between the two classes*.

4.1.3   Important Properties of SVM [1–5]

1. Maximizing the separation between classes is based on a *well-behaved optimization* problem. In the linearly separable case, it is possible to obtain the *globally optimal W*.

2. It can learn *nonlinear boundaries in the input space* by mapping from the input space to a high-dimensional *feature space* and learning a linear boundary in the feature space; such a linear boundary corresponds to a nonlinear boundary in the input space.

3. It employs a suitable *similarity function in the input space* and avoids making expensive computations in the high-dimensional feature space.

4. It combines the training data points to obtain W and use the W for classification.

In its simplest form, the SVM can be used to classify patterns belonging to two classes that are linearly separable.

4.2 Linear SVM [1, 5]

Given the training set $\{X_1, X_2, \ldots, X_n\}$, $X_i \in \mathbb{R}^l$, $i = 1, 2, \ldots, n$

Let the two classes be linearly separable. This means there is a $W \in \mathbb{R}^l$ and a $b \in \mathbb{R}$ satisfying

1. $W^t X_i + b > 0$ $\forall i$ with $y_i = 1$
2. $W^t X_j + b < 0$ $\forall j$ with $y_j = -1$
   We can put these two sets of inequalities together to write
3. $y_k(W^t X_k + b) > 0$ $\forall k$ with $1 \leq k \leq n$.
4. Note that the decision boundary is given by $W^t X + b = 0$. There could be infinitely many possible separating hyperplanes unless we constrain the selection.

4.2.1 Linear Separability

We can study the implication of linear separability as follows:

- Let the training set be $\{(X_1, -1), (X_2, -1), \ldots, (X_{n_-}, -1), (X_{n_-+1}, +1), \ldots, (X_n, +1)\}$
- Note that $W^t X_j + b' = -\varepsilon_j$, where $\varepsilon_j > 0$, $\forall j$ with $y_j = -1$
  Similarly, $W^t X_i + b' = \varepsilon_i$, where $\varepsilon_i > 0$, $\forall i$ with $y_i = 1$
- So, we have $W^t X_j + b' \leq -\varepsilon_-$ where $-\varepsilon_- = \overset{\max}{j} - \varepsilon_j$ and
- We have $W^t X_i + b' \geq \varepsilon_+$ where $\varepsilon_+ = \overset{\min}{i} \varepsilon_i$.
- From these two sets of inequalities, we get
  $W^t X_i + b \leq -\varepsilon$ $\forall i$ with $y_i = -1$ and
  $W^t X_j + b \geq \varepsilon$ $\forall j$ with $y_j = 1$
  where $\varepsilon = \frac{\varepsilon_+ + \varepsilon_-}{2}$ and $b = b' - \frac{\varepsilon_+ - \varepsilon_-}{2}$
- By dividing the two inequalities by ε both sides, we get
  $W_n^t X_i + b_n \leq -1$ $\forall i$ with $y_i = -1$ and
  $W_n^t X_j + b_n \geq 1$ $\forall j$ with $y_j = 1$
  where $W_n = (\frac{w_1}{\varepsilon}, \frac{w_2}{\varepsilon}, \ldots, \frac{w_l}{\varepsilon})^t$ and $b_n = \frac{b}{\varepsilon}$
- Instead of using W_n and b_n, we use W and b, respectively, for the sake of brevity. So, we get the following inequalities
  $W^t X_i + b \leq -1$ $\forall X_i$ such that $y_i = -1$ and
  $W^t X_i + b \geq 1$ $\forall X_i$ such that $y_i = 1$
- Equivalently, we have
  $y_i(W^t X_i + b) \geq 1$, $\forall i$ (because $y_i \in \{-1, +1\}$)

- Note that a pattern X_i with $y_i = 1$ will either lie on the hyperplane $W^t X_i + b = 1$ or it is in the positive side satisfying $W^t X_i + b > 1$.
  Similarly, a pattern X_i with $y_i = -1$ will either fall on the hyperplane $W^t X_i + b = -1$ or it is in the negative side satisfying $W^t X_i + b < -1$.
  So, there is no X_i such that $-1 < W^t X_i + b < 1$ when the classes are linearly separable.
- The hyperplanes $W^t X_i + b = 1$ and $W^t X_i + b = -1$ are called *support planes*.
- The set of training vectors that fall on these support planes can be support vectors.
- When the classes are linearly separable, we can suitably scale W and b to obtain the support planes to satisfy $W^t X_i + b = 1$ and $W^t X_i + b = -1$.
- There is no pattern X_i falling between the two support planes. Further, the two support planes are parallel to each other as shown in Fig. 2.4.

4.2.2 Margin

The distance between the two planes is called the *Margin*. It is possible to show that the margin is a function of W. Training the SVM consists of learning a W that maximizes the margin. So, margin is important in theory.

Consider the point X shown in Fig. 4.1. Let X_{Proj} be the projection of X onto the hyperplane characterized by $g(X) = 0$. Let d be the normal distance between X and the hyperplane, or the distance between X and X_{Proj}, as shown in the figure.

- It is possible to write X in terms of X_{Proj} and d as
  $X = X_{\text{Proj}} + d\frac{W}{||W||}$ because d is the magnitude and the direction is same as that of W. The unit vector in the direction of W is $\frac{W}{||W||}$.

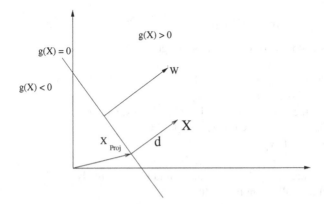

Fig. 4.1 Distance between a point and a hyperplane

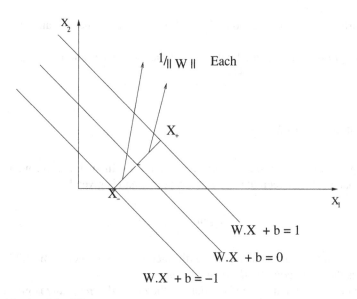

Fig. 4.2 Margin of the SVM

- Observe that

  $g(X) = W^t X + b = W^t(X_{\text{Proj}} + d\frac{W}{||W||}) + b = W^t X_{\text{Proj}} + b + d\frac{W^t W}{||W||} \Rightarrow$

  $g(X) = 0 + d\frac{W^t W}{||W||} = d\frac{W^t W}{||W||}$ because $W^t X_{\text{Proj}} + b = 0$ as X_{Proj} is on $g(X) = 0$.

- $g(X) = d\frac{W^t W}{||W||} \Rightarrow g(X) = d ||W||$.

  So, $d = \frac{g(X)}{||W||}$.

- Hence, the distance between X and the hyperplane $g(X) = 0$ is given by $d = \frac{g(X)}{||W||}$. This result is useful in quantifying the margin.

- Consider Fig. 4.2. We have depicted three parallel lines in the two-dimensional space where $W.X$ is the *dot product* and it is equal to $W^t X$. These are

  1. $W.X + b = -1$ is the support line corresponding to the negative class.
  2. $W.X + b = 0$ which characterizes the decision boundary between the two classes.
  3. $W.X + b = 1$ corresponds to the support plane of the positive class.

- Consider the point X_+ on $W.X + b = W^t X + b = 1$. The normal distance from X_+ to the hyperplane (line in the two-dimensional case) $W.X + b = 0$ ($g(X) = 0$) is given by

  $d = \frac{g(X_+)}{||W||}$; however, $g(X_+) = 1$ because X_+ is on the line (hyperplane in higher dimensions) $g(x) = W.X + b = 1$.

  So, the distance $d = \frac{1}{||W||}$.

- Similarly for the point X_- on $W.X + b = -1$, the normal distance to the line $W.X + b = 0$ is $d = \frac{1}{||W||}$.
- So, *Margin* is characterized by the sum of these distances and is $\text{Margin} = \frac{1}{||W||} + \frac{1}{||W||} = \frac{2}{||W||}$.

4.2.3   Maximum Margin

We are given that the classes are *linearly separable*. In such a case, we have the margin that exists between the two support planes and is given by

$$\text{Margin} = \frac{2}{|| W ||}.$$

The idea is to find out a W that maximizes the margin. Once we get the W, $W^t X + b = 0$ gives us the corresponding *decision boundary*.

More precisely, the decision boundary or the *optimal hyperplane* is given by the solution of the following equivalent optimization problem.

Find $W \in \mathbb{R}^l$, $b \in \mathbb{R}$ to maximize $\frac{2}{W^t W}$ subject to $y_i(W^t X_i + b) \geq 1$, $\forall i$.

Instead of maximizing $\frac{2}{W^t W}$, we can *equivalently minimize* $\frac{W^t W}{2}$ to get

minimize $\frac{1}{2} W^t W$

subject to $y_i(W^t X_i + b) \geq 1$, $i = 1, 2, \ldots, n$

This is an optimization problem with quadratic criterion function $\frac{1}{2} W^t W$ and the constraints are in the form of *linear inequalities* $y_i(W^t X_i + b) \geq 1$.

It is possible to transform the constrained optimization problem into an unconstrained optimization problem using the Lagrangian given by

$$\mathcal{L} = \frac{1}{2} W^t W + \sum_{i=1}^{n} \alpha_i (1 - y_i(W^t X_i + b)).$$

The optimization problem is formulated so that the resulting form is *convex* ensuring globally optimal solution. In this case, the KKT conditions are both *necessary and sufficient*. These are

$$\nabla_W \mathcal{L} = W + \sum_{i=1}^{n} \alpha_i(-y_i)X_i = 0 \Rightarrow W = \sum_{i=1}^{n} \alpha_i y_i X_i.$$

$$\frac{\delta \mathcal{L}}{\delta b} = 0 \Rightarrow \sum_{i=1}^{n} \alpha_i y_i = 0.$$

$$\alpha_i \geq 0 \quad \alpha_i(1 - y_i(W^t X_i + b)) = 0; \quad \text{and} \quad 1 - y_i(W^t X_i + b) \leq 0, \ \forall i.$$

The important properties of the SVM are given by

1. We are given n training patterns and the training set of patterns is
   $\{(X_1, y_1), (X_2, y_2), \ldots, (X_n, y_n)\}$
2. The vector W is given by

$$W = \sum_{i=1}^{n} \alpha_i y_i X_i$$

which means W is a sum of the training patterns which are weighted by the corresponding αs and ys.

We will see later that we need not consider all the training patterns; there will be a small number of patterns with the corresponding αs to be nonzero. We need to consider them only. The other patterns will have their corresponding α values to be 0.

3. The equation

$$\sum_{i=1}^{n} \alpha_i y_i = 0$$

captures the property that

$$\sum_{i=1}^{n_-} \alpha_i = \sum_{i=n_-+1}^{n} \alpha_i.$$

The sum of the αs corresponding to the negative class is equal to that of the positive class. this property is useful in learning W.

4. Another important property, called the *complementary slackness* condition, is given by $\alpha_i(1 - y_i(W^t X_i + b)) = 0$, $\forall i$.

$$\alpha_i > 0 \Rightarrow y_i(W^t X_i + b) = 1$$

which means that if $\alpha_i > 0$, then the corresponding X_i is on a support plane. It is on the positive support plane if $y_i = 1$ else it is on the negative support plane.

4.2.4 An Example

We illustrate the learning of W, b, and αs using a two-dimensional example shown in Fig. 4.3. We have shown two Xs, negative examples, characterized by $(2, 1)^t$, and $(1, 3)^t$ and a O, a positive example, given by $(6, 3)^t$.

Fig. 4.3 Learning W and b from training data

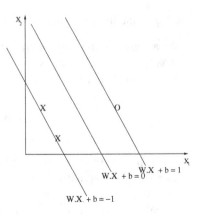

- $(2, 1)^t$ and $(1, 3)^t$ are on the line $W^t X + b = -1$. So, we have

$$2w_1 + w_2 + b = -1$$

and

$$w_1 + 3w_2 + b = -1.$$

- Similarly, $(6, 3)^t$ is on the line $W^t X + b = 1$. So, we get

$$6w_1 + 3w_2 + b = 1.$$

- Solving the three equations, we get $w_1 = \frac{2}{5}$, $w_2 = \frac{1}{5}$, and $b = -2$.
- Note that $(4, 2)^t$ is on the boundary as $W^t(4, 2)^t + b = (\frac{2}{5}, \frac{1}{5})(4, 2)^t - 2 = 0$. Similarly, $(7, 1)^t$ is in the positive class and $(2, 0)^t$ is in the negative class.
- Further,

$$\sum_i \alpha_i y_i = 0 \Rightarrow -\alpha_1 - \alpha_2 + \alpha_3 = 0 \Rightarrow \alpha_3 = \alpha_1 + \alpha_2.$$

- Also

$$W = (\frac{2}{5}, \frac{1}{5})^t = -\alpha_1(1, 3)^t - \alpha_2(2, 1)^t + \alpha_3(6, 3)^t \Rightarrow$$

$$\alpha_1 = 0; \ \alpha_2 = \alpha_3 = \frac{1}{10}.$$

Note that $\alpha_1 = 0$. So, it is possible that αs corresponding to some of the patterns on the support planes could be 0.

4.3   Dual Problem

If we substitute

$$W = \sum_{i=1}^{n} \alpha_i y_i X_i$$

in the Lagrangian $\mathscr{L}$, we have

$$\mathscr{L} = \frac{1}{2} \sum_{i=1}^{n} \alpha_i y_i X_i^t \sum_{j=1}^{n} \alpha_j y_j X_j + \sum_{i=1}^{n} \alpha_i \left(1 - y_i (\sum_{j=1}^{n} \alpha_j y_j X_j^t X_i + b) \right)$$

By simplifying further, we get

$$\mathscr{L} = \sum_{i=1}^{n} \alpha_i - \frac{1}{2} \sum_{i=1}^{n} \sum_{j=1}^{n} \alpha_i \alpha_j y_i y_j X_i^t X_j + b \sum_{i=1}^{n} \alpha_i y_i$$

By noting that

$$\sum_{i=1}^{n} \alpha_i y_i = 0,$$

we get

$$\mathscr{L} = \sum_{i=1}^{n} \alpha_i - \frac{1}{2} \sum_{i=1}^{n} \sum_{j=1}^{n} \alpha_i \alpha_j y_i y_j X_i^t X_j .$$

This is the *dual problem* and it is in terms of αs only. We use $\mathscr{L}_D$ for the dual and it is

$$\mathscr{L}_D(\alpha) = \sum_{i=1}^{n} \alpha_i - \frac{1}{2} \sum_{i=1}^{n} \sum_{j=1}^{n} \alpha_i \alpha_j y_i y_j X_i^t X_j$$

such that $\alpha_i \geq 0, \ \forall i$ and $\sum_{i=1}^{n} \alpha_i y_i = 0$

1. This is a *convex optimization problem*. It is possible to obtain α vector corresponding to the *global optimum*.
2. The vector $W = \sum_{i=1}^{n} \alpha_i y_i X_i$. So, optimization is over $\mathbb{R}^n$ *irrespective of the dimension of* X_i.
3. Many of the α_i are 0. *Support Vectors (SVs)* are the X_is corresponding to the nonzero α_is.
4. Let, $S = \{X_i | \alpha_i > 0\}$ be the set of SVs.

a. By *complementary slackness condition*, $X_i \in S \Rightarrow \alpha_i > 0 \Rightarrow$
   $y_i(W^t X_i + b) = 1 \Rightarrow$.
   X_i is the closest to the decision boundary.
b. We have $W = \sum_i \alpha_i y_i X_i = \sum_{X_i \in S} \alpha_i y_i X_i$.
   Optimal W is a *linear combination of the support vectors*.
c. $b = y_j - W^t X_j$, where j is such that $\alpha_j > 0$.
d. Thus, both W and b are determined by α_j, $j = 1, 2, \ldots, n$.
e. We can solve the dual optimization problem to obtain the optimal values of
   α_is. We can use the αs to get optimal values of both W and b.
f. Typically we would like to classify a new pattern Z based on the sign of
   $W^t X + b$.
   Equivalently, by using $W = \sum_i \alpha_i y_i X_i$, we can classify a pattern Z based
   on the sign of $b + \sum_{X_i \in S} \alpha_i y_i X_i^t Z$. We do not need to use W explicitly.

4.3.1   An Example

Let us consider the example data shown in Fig. 4.4. There are five points. These are

- **Negative Class**: $(2, 0)^t$, $(2, 1)^t$, $(1, 3)^t$
- **Positive Class**: $(6, 3)^t$, $(8, 2)^t$

We have seen earlier that $W = (\frac{2}{5}, \frac{1}{5})^t$ and $b = -2$ for the patterns $(2, 1)^t$, $(1, 3)^t$,
$(6, 3)^t$, first two from the negative class and the third from the positive class.
  The α values are $\alpha_1 = 0$, $\alpha_2 = \alpha_3 = \frac{1}{10}$.
  The remaining two patterns are such that the corresponding αs are 0.

1. $(2, 0)^t$ is from class -1. Based on the complementary slackness condition, we
   have $\alpha(1 - y(W^t X + b)) = 0$. Here, $y = -1$, $W = (\frac{2}{5}, \frac{1}{5})^t$, $X = (2, 0)^t$, and
   $b = -2$. So, $\alpha = 0$ because $1 - y(W^t X + b) = -\frac{1}{5} \neq 0$.

Fig. 4.4 α values

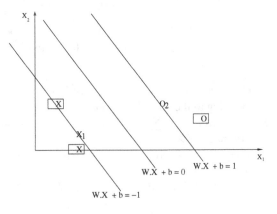

2. $(8, 2)^t$ is from class $+1$. Here, $\alpha = 0$ because $1 - y(W^t X + b) = -\frac{3}{5}$.
3. So, the SVs are $X_1 = (2, 1)^t$, $O_2 = (6, 3)^t$ and both have the same α value of $\frac{1}{10}$. The α values corresponding to the other three patterns are 0.
4. The points which are not support vectors or equivalently points with zero α value are indicated using a rectangular box around them in Fig. 4.4.

4.4 Multiclass Problems [2]

Classifiers like perceptron and SVM are based on linear discriminants and are ideally suited for two-class problems or binary classification problems. So, when the training data is from C (> 2) classes, then we need to build a *multiclass classifier* from a collection of binary classifiers. Some of the well-known possibilities are

1. Consider a *pair of classes* at a time; there are $\frac{C(C-1)}{2}$ such pairs. Learn a linear discriminant function for each pair of classes.
   Consider Fig. 4.5.

   These decisions are combined to arrive at the class label among the three classes $C_1, C_2,$ and C_3. Note that there are three binary classifiers as shown in the figure. A problem is the ambiguous region marked in the middle. It is difficult to classify a point in this region.
2. For class C_i let the complementary region be

$$\bar{C}_i = \bigcup_{j=1, j \neq i}^{C} C_j$$

   Learn a linear discriminant function to classify to C_i or $\bar{C}_i$ for each i. Combine these binary classifiers to classify a pattern.

Fig. 4.5 Multiclass classification

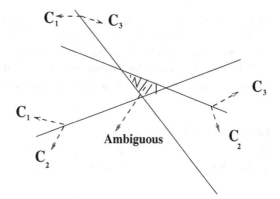

Fig. 4.6 Multiclass
classification

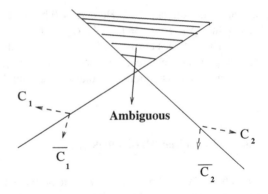

Consider Fig. 4.6.

Note that even in this case, there is a region that is ambiguous as shown in the figure.

4.5   Experimental Results

Here, we considered Iris Setosa and Iris Versicolour classes which are linearly separable. We used the two features *sepal length* and *petal length* in building and testing the classifiers. We have used 60 patterns for training and the remaining 40 for testing. Both the perceptron and linear SVM classifiers have given us 100 % accuracy on the test data set. The weight vectors learnt are given in Table 4.1. Here, W_p is the weight vector learnt using Perceptron and W_s is the weight vector obtained using SVM.

4.5.1   Results on Multiclass Classification

SVM and Perceptron are inherently two-class classifiers. We use the traditional way of one-against-rest method to perform multiclass classification. *Weka*, a popular suite of machine learning software is used in realizing this.

We consider two well-known machine learning data sets: Iris and Pendigits. The number of instances, attributes, and the results are listed below. The data set is split

Table 4.1 Directions of W_p and W_s

W_p	W_s	Cosine (W_p, W_s)
$(2, 3.4, -9.1)^t$	$(-1, 0.9827, -1.96)^t$	0.80

Table 4.2 Results on iris dataset with three classes

No of training training patterns	No of test test patterns	Number of correctly classified patterns	Accuracy (Percentage)
75	75	71	94.67
83	67	63	94.03
90	60	57	95
99	51	49	96.08
105	45	43	95.56
110	40	38	95
113	37	35	94.60
117	33	32	96.67
120	30	29	96.67

into train and test sets, and fed into the multiclass classifier of Weka. Several iterations are carried out with different train-test percentage splits. Finally the Mean and the Standard Deviation are calculated. Also, we have provided the results for multiclass classifier using a tenfold cross validation.

We give below the details of our experiments.

1. **Iris Dataset**

   Number of Classes = 3

   Number of Data Points = 150

   Number of features = 5

   Out of the 5 features, 4 of them are *sepal length, sepal width, petal length,* and *petal width.* The *fifth* feature is a dependent feature; it is the *class label* which can assume one of three values corresponding to the 3 classes, Setosa, Versicolour, and Viriginica. We give the results in Table 4.2.

   By using tenfold cross validation, we obtained an accuracy of 96 %.

2. **Pendigits Dataset**

   Number of Classes = 10

   Number of Data Points = 10992

   Number of Features = 17

   Out of the 17 features, the 17th feature is the class label assuming one of 10 values corresponding the digit that is represented by 16 features. We have used Weka software that is described in the book by Witten, Frank and Hall, the details of which are provided in the references. We give results in Table 4.3.

   Using tenfold cross validation, we could classify with an accuracy of 93.52 %

Table 4.3 Results on pendigit dataset with ten classes

No of training training patterns	No of test test patterns	Number of correctly classified patterns	Accuracy (Percentage)
5496	5496	5145	93.61
5946	4946	4644	93.90
6695	4397	4124	93.79
7255	3737	3499	93.63
7694	3298	3101	94.02
8244	2748	2579	93.85
8574	2418	2270	93.88
8794	2198	2065	93.95

4.6   Summary

Classification based on SVMs is popular and is being used in a variety of applications. It is good to understand why it works and also its shortcomings. Some of the important features are

1. Both SVM and perceptron are *linear classifiers*.
2. It is possible to view the linear classifier to have the form $W^t X + b$. The training patterns are used to *learn* W and b.
3. In both the SVM and perceptron, the W vector may be viewed as a *linear combination* of the training patterns.

   a. In perceptron the iterations converge to a W_{k+1}, a correct weight vector, and it is

   $$W_{k+1} = \sum_{i=1}^{k} X^i,$$

   where X^k is misclassified by W_k.
   b. In SVM, the weight vector W is given by

   $$W = \sum_{X_i \in S} \alpha_i y_i X_i,$$

   where only *support vectors* matter.

4. Consider the data shown in Fig. 4.7.

   Here, there are two points each from the two classes as given by

   a. Negative Class: $X_1 = (1, 1)^t$, $X_2 = (1, 6)^t$

Fig. 4.7 Different support
vector Sets

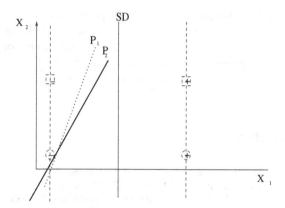

b. Positive Class: $X_3 = (5, 1)'$, $X_4 = (5, 6)'i$

The W and b vectors given by Perceptron and SVM can be obtained as follows:

a. If we use the order X_1, X_2, X_3, X_4, to compute the augmented W using
   the perceptron learning algorithm, we get $W = (5, -3)'$ and $b = -3$, the
   corresponding decision is indicated using P_1 in the figure.
b. If we use the order X_3, X_4, X_2, X_1, we get $W = (12, -9)'$, and $b = -4$
   and we show the corresponding decision boundary using P_2 in the figure.
c. Using the SVM, we get two possible support vector sets. They are
   - The support vectors are $X_1 = (1, 1)'$ and $X_3 = (5, 1)'$. Because of $\sum_i \alpha_i$
     $y_i = 0$, we get
     $-\alpha_1 + \alpha_3 = 0 \Rightarrow \alpha_1 = \alpha_3 = \alpha$. So,
     $W = \alpha[(5, 1)' - (1, 1)'] = (4\alpha, 0)$.
     Also, because $(1, 1)'$ is on the negative support line, we get
     $w_1 + w_2 + b = -1$.
     Similarly, for $(5, 1)'$ which is on the positive support line, we have
     $5w_1 + w_2 + b = 1$.
     From these two equations, we get $w_1 = \frac{1}{2}$ and $w_2 = 0$. So, $W = (4\alpha, 0)' = $
     $(\frac{1}{2}, 0)' \Rightarrow \alpha = \frac{1}{8}$.
     From these, we get $b = -\frac{3}{2}$. So, the decision boundary, SD, is character-
     ized by $\frac{x_1}{2} - \frac{3}{2} = 0$ as shown in the figure.
   - The other possibility is to have $X_2 = (1, 6)'$ and $X_4 = (5, 6)'$. Here also
     we get $W = (\frac{1}{2}, 0)'$ and $b = -\frac{3}{2}$. Again the decision boundary is given
     by SD. In both the cases, W is orthogonal to SD.
   Even though both the Support Vector sets are different, we get the same W.
   So, in the case of the SVM also we can have multiple solutions, in terms of
   the SV sets. However, the W vector is the same.

5. Linear Support Vector Machine is a simple linear classifier. It is popularly used
   in linearly separable cases.

6. It is also used in classifying high-dimensional datasets even if the classes are not linearly separable. Some of the popular applications are **text classification** and classification of nodes and edges in **social networks**.
7. Experimental results on Iris data do not show much difference between Perceptron and Linear SVM in terms of accuracy.

References

1. Fan, R.-E., Chang, K.-W., Hsieh, C.-J, Wang, X.-R., Lin, C.-J.: LIBLINEAR: a library for large linear classification. JMLR **9**, 1871–1874 (2008)
2. Hsu, C.W., Lin, C.-J.: A comparison of methods for multiclass support vector machines. IEEE Trans. Neural Networks **13**(2), 415–425 (2002)
3. Manning, C.D., Raghavan, P., Schütze, H.: Introduction to Information Retrieval. Cambridge University Press (2008)
4. Rifkin, R.M.: Multiclass Classification, Lecture Notes, Spring08. MIT, USA (2008)
5. Witten, I.H., Frank, E., Hall, M.A.: Data Mining, 3rd edn. Morgan Kauffmann (2011)

Chapter 5
Kernel-Based SVM

Abstract Kernel Support Vector Machine (SVM) is useful to deal with nonlinear classification based on a linear discriminant function in a high-dimensional (kernel) space. Linear SVM is popularly used in applications involving high-dimensional spaces. However, in low-dimensional spaces, kernel SVM is a popular nonlinear classifier. It employs *kernel trick* which permits us to work in the *input space* instead of dealing with a potentially high-dimensional, even theoretically infinite dimensional, kernel (feature) space. Also kernel trick has become so popular that it is used in a variety of other pattern recognition and machine learning algorithms.

Keywords Kernel trick · Soft margin formulation · Kernel function · Nonlinear decision boundary

5.1 Introduction

In the last two chapters, we have learnt the classifiers based on the assumption that the data is linearly separable. However, we need to ask

5.1.1 *What Happens if the Data Is Not Linearly Separable? [2–4, 6]*

The immediate observations are

1. There are no feasible W and b such that $W^t X + b = -1$ if X is from the negative class and $W^t X + b = 1$ if X belongs to the positive class.
2. There is no *margin*. So, maximizing the margin does not make sense. It is possible that a good number of *practical problems* fall under this category.
3. So, we cannot find the optimal hyperplane using the formulation considered in the previous chapter when the data is not linearly separable.
4. Instead we *create margin* and optimize; in a sense we consider *ignoring* some points to create the margin.

© The Author(s) 2016
M.N. Murty and R. Raghava, *Support Vector Machines and Perceptrons*,
SpringerBriefs in Computer Science, DOI 10.1007/978-3-319-41063-0_5

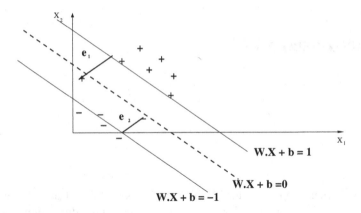

Fig. 5.1 Error in classification

We consider the corresponding formulation next.

5.1.2 Error in Classification

- Consider the two-dimensional data points shown in Fig. 5.1.
- A pattern X is assigned to class $+$ if $W^t X + b = W \cdot X + b > 0$. Similarly, X is assigned to class $-$ if $W^t X + b = W \cdot X + b < 0$.
- A positive pattern is erroneously classified with an error of e_1 that is greater than 1.
- A negative pattern is associated with an error of e_2 (<1) as shown in the figure. There is no misclassification here.
- We would like to minimize such errors. So, we include a term corresponding to the sum of such errors in the criterion function. Hence, the optimization problem boils down to

$$Minimize \; \frac{1}{2} \parallel W \parallel^2 + C \sum_{i=1}^{n} e_i$$

- There is no error in classifying X_i if $e_i = 0$. Also e_i cannot be negative. So, $e_i \geq 0 \; \forall i$.
- Similarly, the constraints now can be relaxed as

$$W^t X_i + b \geq -1 + e_i, \;\; if \;\; y_i = -1 \;\; and \; W^t X_i + b \leq 1 - e_i, \;\; if \;\; y_i = 1$$

Note that the introduction of error e_i for the pattern $X_i \in C_+$ ensures that the corresponding constraint is satisfied. There are three possibilities:

1. The error $e_i = 0$. In this case X_i is on the support plane ($W^t X_i + b = 1$) and so is correctly classified.

2. If $e_i < 1$, then $W^t X_i + b \geq 1 - e_i > 0$. So, X_i will be correctly classified. However, X_i is on the margin.
3. If $e_i \geq 1$, then $W^t X_i + b \leq 0$. So, X_i will be misclassified.

A similar analysis can be made for patterns in C_-.

Note that irrespective of whether $X_i \in C_+$ ($y_i = 1$) or $X_i \in C_-$ ($y_i = -1$), we have

$y_i(W^t X_i + b) \geq 1 - e_i$ and also $e_i \geq 0$ $\forall i$.

5.2 Soft Margin Formulation [2]

Associating Lagrange variables with the constraints, we get the Lagrangian

$$\mathcal{L} = \frac{1}{2} W^t W + C \sum_{i=1}^{n} e_i + \sum_{i=1}^{n} \alpha_i(1 - e_i - y_i(W^t X_i + b)) - \sum_{i=1}^{n} \mu_i e_i$$

where

μ_i is the Lagrange variable associated with the constraint $e_i \geq 0$ or equivalently, $-e_i \leq 0$.

Similarly α_i is the Lagrange variable associated with the constraint $y_i(W^t X_i + b) \geq 1 - e_i$ or equivalently with $1 - e_i - y_i(W^t X_i + b) \leq 0$.

5.2.1 The Solution

1. By taking the gradient of $\mathcal{L}$ with respect to W and equating to 0, we get

$$W = \sum_{i=1}^{n} \alpha_i y_i X_i$$

which is the same as the expression obtained in the hard margin formulation studied in the previous chapter.
2. By differentiating $\mathcal{L}$ by b and equating to 0, we get

$$\sum_{i=1}^{n} \alpha_i y_i = 0.$$

Again this is the same as the one seen in the previous chapter.
3. $\frac{\delta \mathcal{L}}{\delta e_i} = 0 \Rightarrow \alpha_i + \mu_i = C$
4. $\alpha_i \geq 0$; $\mu_i \geq 0$; and $e_i \geq 0$ $\forall i$.
5. Also, $\alpha_i(1 - e_i - y_i(W^t X_i + b)) = 0$ and $\mu_i e_i = 0$. because of the constraints.

5.2.2   *Computing b*

1. Consider an X_i for which α_i is such that $0 < \alpha_i < C$, then $\mu_i > 0$ because $\alpha_i + \mu_i = C$.
2. If $\mu_i > 0$, then $e_i = 0$ as $\mu_i e_i = 0$ for all i.
3. Based on these values of α_i and μ_i, we get $(1 - y_i(W^t X_i + b)) = 0$ or $b = y_i - W^t X_i$ as $\alpha_i(1 - e_i - y_i(W^t X_i + b)) = 0$.

5.2.3   *Difference Between the Soft and Hard Margin Formulations*

1. In the soft formulation $\alpha_i + \mu_i = C$ and we require to ensure that $\mu_i \geq 0$. So, we need $0 \leq \alpha_i \leq C$.
2. So, in the soft formulation, α_i is bounded above by C.
3. However, in the hard formulation, α_i is not bounded above. Recall that $\alpha_i \geq 0$.
4. If we let $C \to \infty$, then there is no difference between the two. This simply means that we cannot tolerate any error; every $e_i = 0$. As a consequence,

$$\sum_{i=1}^{n} e_i = 0$$

and so the error term

$$C \sum_{i=1}^{n} e_i = 0.$$

5.3   Similarity Between SVM and Perceptron

1. In the case of Perceptron, the optimal value of W given by W_P is

$$W_P = minarg_W \ - \sum_{i=1}^{n} y_i(W^t X_i + b_P).$$

Here, the criterion function corresponds to sum of the errors and it is minimized when none of the patterns is misclassified by W_P or when the two classes are linearly separable. Note that b_P is the bias term in perceptron. So, perceptron may be viewed as minimizing some error.

In the case of soft margin formulation of SVM, the criterion function is

$$\frac{1}{2} \| W_S \|^2 + C \sum_{i=1}^{n} e_i$$

where W_S is the optimal weight vector corresponding to the soft margin SVM and e_i is the error corresponding to X_i. C is a hyper parameter to be specified by the user. If C is chosen such that

$$\sum_{i=1}^{n} e_i$$

is larger than

$$\frac{1}{2} \| W_S \|^2,$$

then SVM is also minimizing some error.

2. The constraints in Perceptron may be viewed as
$y_i(W_P^t X_i + b_P) > 0$
whereas for the soft margin SVM, they are
$y_i(W_S^t X_i + b_S) \geq 1 - e_i$.
If $e_i < 1$, then the constraints have some similarity and are respectively
$y_i(W_P^t X_i + b_P) > 0$
and
$y_i(W_S^t X_i + b_S) > 0$.

3. Note that in the case of perceptron, the weight vector W_P without augmentation and multiplication with the class label value (of $+1$ or -1) is given by

$$W_P = \sum_{i=1}^{n} \beta_i y_i X_i$$

where $y_i = 1$ if $X_i \in C_+$ and $y_i = -1$ if $X_i \in C_-$. Further, β_i is a nonnegative integer and corresponds to the number of times X_i is misclassified before the percetron learning algorithm converges to the correct weight vector W_P. Similarly W_S corresponding to the SVM with soft or hard margin is

$$W_S = \sum_{i=1}^{n} \alpha_i y_i X_i$$

where α_i is nonnegative real number.

4. In the soft margin formulation, selection of C can be crucial. There are applications where the value of C could be very large. In such cases, it could be similar to the hard margin formulation.

   It could the case that some values of C might make the SVM work like the perceptron.

5.4 Nonlinear Decision Boundary [1, 6]

So far we have considered large-margin SVM classifiers. It is possible, in some applications, that the best linear boundary may not be adequate. In such a case, it is useful to consider a nonlinear boundary.

One possibility is to *map X_i to a higher dimensional space and obtain a linear decision boundary in the new space.*

Consider the example data shown in Fig. 5.2.

The terminology is

1. **Input Space**: Space of points X_i.
2. **Feature Space**: The space of $\phi(X_i)$ after transformation. In general, we can have $\phi : \mathbb{R}^m \to \mathbb{R}^{\hat{m}}$.
3. Corresponding to the pair (X_i, y_i) in the input space, we have the pair (Z_i, y_i) in the feature space. Here, $Z_i = \phi(X_i)$.
4. The problem boils down to finding the optimal hyperplane in the feature space by solving the dual problem where $Z_i^t Z_j$ replaces $X_i^t X_j$.

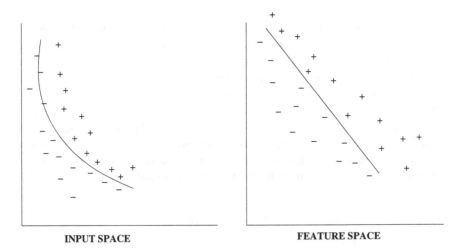

| INPUT SPACE | FEATURE SPACE |

Fig. 5.2 Nonlinear decision boundary

5.4.1 Why Transformed Space?

1. A nonlinear boundary in the input space could be captured using a linear decision boundary in the feature space.
2. If the transformation is appropriate, then it may be possible to realize a simple linear classifier in the feature space that can capture the required nonlinearity in the input space.
3. Typically feature space is of higher dimensionality compared to the input space.
4. *Kernel trick* is employed to perform the computation in the input space to obtain the solution in the feature space.
5. It is important to note that we need not know the mapping ϕ explicitly; in theory, it could be mapping a point X in the input space to the feature space that is *infinite dimensional*.

5.4.2 Kernel Trick

1. Typically, a kernel function, $K : \mathbb{R}^m \times \mathbb{R}^m \to \mathbb{R}$ where $K(X_i, X_j) = \phi(X_i)^t \phi(X_j)$ is used.
2. In this setting, we have

$$W = \sum_{X_i \in S} \alpha_i y_i \phi(X_i)$$

where S is the set of support vectors.
3. Given a test pattern X, we can classify it based on $W^t \phi(X) + b$ which is given by

$$\sum_{X_i \in S} \alpha_i y_i \phi(X_i)^t \phi(X) + b$$

4. Note that b is obtained by

$$b = y_p - W^t \phi(X_p) = y_p - \sum_{X_i \in S} \alpha_i y_i \phi(X_i)^t \phi(X_p)$$

5. So, kernel computation involves computing dot product between vectors in the *feature space*.
6. In other words, kernel computes some kind of similarity between the test pattern and each of the support vectors in the feature space.
7. We have $K(X_i, X_j) = \phi(X_i)^t \phi(X_j)$. So, the dot product in the *feature space* is computed using the kernel computation in the *input space*.

5.4.3 An Example

1. Let us consider using the kernel $K(X_i, X_j) = 1 + X_i^t X_j$.
2. Let $X_i = (x_{i1}, x_{i2}, \ldots, x_{il})^t$ and $X_j = (x_{j1}, x_{j2}, \ldots, x_{jl})^t$.
3. $K(X_i, X_j) = 1 + x_{i1}x_{j1} + x_{i2}x_{j2} + \cdots + x_{il}x_{jl}$.
4. Consider the mapping $\phi(X) = (1, x_1, x_2, \ldots, x_l)^t$.
5. Now $\phi(X_i)^t \phi(X_j) = 1 + x_{i1}x_{j1} + x_{i2}x_{j2} + \cdots + x_{il}x_{jl}$. Note that $K(X_i, X_j) = \phi(X_i)^t \phi(X_j)$.
6. So, we can compute $K(X_i, X_j)$ in the input space as shown in step 3. It is equivalent to the computation in the feature space as shown in step 5.
7. This is a simple example dealing with a linear kernel. We can show such computational possibilities for a variety of other types of kernels also.

5.4.4 Example Kernel Functions

1. *Linear Kernel*: We have seen an example of this type of kernel already.
2. *Polynomial Kernel*:

$$K(X_i, X_j) = (1 + X_i^t X_j)^P$$

is a polynomial kernel of degree P.
3. *Gaussian Kernel*:

$$K(X_i, X_j) = e^{-\frac{||X_i - X_j||^2}{\sigma^2}}.$$

Note that the exponential function can have infinite terms in its expansion and so the corresponding ϕ could be mapping an X into an infinite dimensional space.

5.5 Success of SVM [2, 5]

1. Theoretical issues:

   a. The success of SVM may be attributed, in theory, to the optimization problem posed as a convex quadratic optimization problem. The criterion function has a single optimum.
   b. The notion of *maximum margin* may not directly lead to an intuitively acceptable classifier.
   c. There are other theoretical results based on, for example, the *Vapnik–Chervonenkis (VC) dimension* which have not been so useful in characterizing the SVMs in a precise manner.
   d. Kernel or similarity function is a powerful idea. The associated notion called *kernel trick* has been exploited in a wide variety of pattern recognition tasks

including the Nearest Neighbor classifier, K-Means algorithm, Fisher's linear discriminant, and Principal Component analysis.

e. Typically classifiers fail to do well in high-dimensional spaces. In order to avoid *overfitting*, they require a larger number of training patterns. Here again the kernel trick that permits us to deal with the low-dimensional *input space* instead of the high-dimensional *feature space*.

2. Practical issues:

a. *Soft margin formulation* is the most practical aspect associated with SVMs.

b. In most of the high-dimensional applications including text classification linear SVMs are popularly used even when the classes are not linearly separable. In such cases the problem is specified using the soft margin formulation; hence tuning the value of the parameter C is important.

c. It is difficult to select the kernel function and tune the parameters associated with the kernel, especially in high-dimensional spaces.

d. When the classes are not linearly separable, the notion of margin and maximizing its value are not meaningful. Margin itself does not exist when the classes are not linearly separable.

e. Another important observation is that the support vectors are *boundary patterns* or *noisy patterns*. Most of the other classifiers work using the core part of the data. Whereas the SVM is distinct; it works using the set of support vectors that lie on the boundaries.

f. One of the important practical issues is the tuning of the parameters to run the SVM successfully.

g. The popularity of SVMs, in practice, could be attributed to the software packages that have been developed and used frequently by users. Notable contributions, in this direction, are LIBSVM and LIBLINEAR.

5.6 Experimental Results

We have used both the Iris dataset and the Digit recognition data set in these experiments. We provide the details below.

5.6.1 *Iris Versicolour and Iris Virginica*

We have considered the two classes that are not linearly separable in the Iris dataset. We used the features *Sepal Length* and *Petal length*. The weight vectors are shown in Table 5.1.

We stop the iterations on the update of W of perceptron when a small number of errors are made. The W_P and W_S vectors corresponding respectively to perceptron and SVM are highly similar.

Table 5.1 Results comparing perceptron and SVM

W_p	W_s	Cosine (W_p, W_s)
$(0, 14, -26)^t$	$(-1, 1.56, -2)^t$	0.92

5.6.2 Handwritten Digit Classification

Here we have used the perceptron and Kernel SVM in classifying the handwritten digit data. We have obtained the following results:

- Perceptron: 100 % accuracy
- Kernel SVM: 99.85 % accuracy

5.6.3 Multiclass Classification with Varying Values of the Parameter C

We have considered three additional benchmark datasets to study the behavior of the soft margin SVM on multiclass problems and simultaneously testing its behavior with changes in the parameter C. The details of the datasets considered are given in Table 5.2.

We varied the value of C between 1 and 128 in all the cases. We have used **one versus the rest** classification scheme in all the three cases. Please see the lecture notes of Rifkin that is given as a reference at the end. We have used the Weka software for running the SVM classifier. The classification accuracies obtained using the linear SVM are shown in Table 5.3.

Some of the observations based on results shown in the Tables 5.2 and 5.3 are:

1. On different datasets the SVM classifier has given different classification accuracies. However, the best accuracy in all the three cases is above 95 %.
2. In the case of *Pen digits* dataset, the best accuracy is obtained at a larger value of C. However, the dimensionality of the data is only 16. The *USPS* dataset has a dimensionality of 256 followed by the *DNA* dataset that has a dimensionality of 180.
3. Note that as the value of C increases, the softness of the classifier decreases.
4. An important observation is that on different datasets, different C values, have resulted in the best accuracy.

Table 5.2 Details of the benchmark datasets

Dataset	Number of classes	Number of features	Number of training patterns	Number of test patterns
USPS	10	256	7291	2007
DNA	3	180	2000	1186
Pen digits	10	16	794	3498

Table 5.3 Classification accuracy: variation with the value of C

Value of parameter C	% Accuracy (USPS)	% Accuracy (DNA)	% Accuracy (Pen digits)
1	93.92	94.52	93.71
2	94.47	95.11	95.57
4	94.87	95.11	96.83
8	95.07	**95.45**	97.40
16	95.17	95.36	97.46
32	**95.27**	95.36	97.54
64	95.07	95.36	97.71
128	94.87	95.36	**97.86**

5.7 Summary

In this chapter, we have discussed the kernel SVM which can deal with nonlinearly separable problems. Some of the important aspects of this classifier are

1. The kernel functions permit us to map patterns in the input space to a potentially infinite dimensional feature space.
2. Even though the feature space is high-dimensional it is possible to exploit the kernel properties to do the computations in the low-dimensional input space.
3. Kernel function is a kind of similarity function that considers a pair of patterns at a time and computes the similarity between the two patterns.
4. Kernel trick is not only used in SVM-based classification, but also in a variety of other applications.
5. Tuning the parameters when Gaussian kernels are used can be a challenge in high-dimensional spaces.
6. The experimental results on some benchmark datasets show that the best C value can be different for different datasets. So tuning C is a challenge in general even while using the Linear SVM.

References

1. Asharaf, S., Murty, M.N., Shevade, S.K.: Multiclass core vector machine. In: Proceedings of International Conference on Machine Learning, 20–24 June 2007, pp. 41–48. Corvallis, Oregon, USA (2007)
2. Burges, C.J.C.: A tutorial on support vector machines for pattern recognition. Data Min. Knowl. Disc. 2(2), 121–167 (1998)
3. Rifkin, R.M.: Multiclass Classification, Lecture Notes, Spring08. MIT, USA (2008)
4. Schölkopf, B., Smola, A.J.: Learning with Kernels. MIT Press (2001)
5. Vishwanathan, S.V.N., Smola, A.J., Murty, M.N.: Simple SVM. In: Proceedings of International Conference on Machine Learning, 21–24 August 2003, pp. 760–767. Washington, D.C., USA (2003)
6. Witten, I.H., Frank, E., Hall, M.A.: Data Mining. Third Edition, Morgan Kauffmann (2011)

Chapter 6
Application to Social Networks

Abstract Social and information networks are playing an important role in several applications. One of important problems here is classification of entities in the networks. In this chapter, we discuss several notions associated with social networks and the role of linear classifiers.

Keywords Social network · Community detection · Link prediction · Learning · Similarity function · Supervised learning

6.1 Introduction

We discuss some of the issues related to representation of networks using graphs after introducing some basic terms.

6.1.1 What Is a Network?

A **network** is a structure made up of a set of nodes and possible links between nodes. To simplify the discussion, we assume that there is at most one link between any pair of nodes. There are several applications where networks are commonly encountered and analyzed. Some of the well-known networks are *internet* and *electrical network*. For example, in the case of internet, a web page is a node in the network and a *hyperlink* from a web page to another web page is a link.

6.1.2 How Do We Represent It?

Typically, such a network is represented by a simple graph. We illustrate it with the help of an example network shown in Fig. 6.1.

© The Author(s) 2016
M.N. Murty and R. Raghava, *Support Vector Machines and Perceptrons*,
SpringerBriefs in Computer Science, DOI 10.1007/978-3-319-41063-0_6

Fig. 6.1 Example network

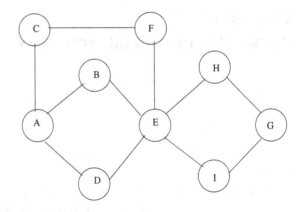

Some of the observations are:

- There are *nine nodes* in the example network; they are labeled using *A* to *I*.
- Between some pairs of nodes, *there is a link*; for example, node pairs *A*, *B* and *G*, *I*.
- There is *no link* between some pairs of nodes; for example, node pairs *A*, *E* and *H*, *I*.
- The graph shown in Fig. 6.1 is *undirected*. For example, the link between *A* and *B* may be represented either by ⟨*A*, *B*⟩ or ⟨*B*, *A*⟩. Both are same. Such a representation conveys the notion of *association* between the two nodes. For example, *A is a friend of B* is the same as *B is a friend of A*. Such relations that are symmetric can be captured by undirected edges/links in the graph.
- There are three *simple paths* between *A* and *E*. These are

  1. Path *A*, *B*, *E*; here *B* is linked to both *A* and *E*. So, *B* is a *common neighbor* of *A* and *E*.
  2. Path *A*, *D*, *E*; here *D* is linked to both *A* and *E*. So, *D* is a common neighbor of *A* and *E*.
  3. Path *A*, *C*, *F*, *E*; here neither *C* nor *F* are linked to both *A* and *E*. So, *neither C nor F* is a common neighbor of *A* and *E*.
  4. There is a notion of *degree* associated with each node. The *degree of a node* is the number of links associated with the node. The nodes in the example graph have the following degree profile:

- *Nodes of degree 2*: the nodes labeled *B*, *C*, *D*, *F*, *G*, *H*, *I* are of degree 2.
- *Nodes of degree 3*: observe that *A* has degree 3.
- *Nodes of degree 5*: node *E* has degree 5.

So, networks are *represented using graphs*. Further, graphs are popularly represented on the machine using two popular schemes. These are based on either *adjacency matrix* or *adjacency list*.

Table 6.1 Adjacency matrix for the graph in Fig. 6.1

Node /Node	A	B	C	D	E	F	G	H	I
A	0	1	1	1	0	0	0	0	0
B	1	0	0	0	1	0	0	0	0
C	1	0	0	0	0	1	0	0	0
D	1	0	0	0	1	0	0	0	0
E	0	1	0	1	0	1	0	1	1
F	0	0	1	0	1	0	0	0	0
G	0	0	0	0	0	0	0	1	1
H	0	0	0	0	1	0	1	0	0
I	0	0	0	0	1	0	1	0	0

- **Adjacency Matrix (A_G):**

  - The adjacency matrix of a graph is a *square matrix* of size $n \times n$ where n is the number of nodes in the graph.
  - The ijth entry in the matrix is 1 if there is a link between nodes i and j; if there is no link then the entry is 0.
  - If the *graph is undirected*, then the adjacency matrix is *symmetric* also. The adjacency matrix corresponding to the undirected graph in Fig. 6.1 is given in Table 6.1.
  - **Number of Paths of Length 2:**
    The adjacency matrix A_G of a graph G characterizes edges or paths of length 1. We get paths of length 2 by considering the matrix $A_G^2 (A_G \times A_G)$. The matrix A_G^2 for the example graph in Fig. 6.1 is given in Table 6.2
  - Note that in A_G^2 the diagonal entries correspond to the degrees of the respective nodes.

Table 6.2 Square of the adjacency matrix for the graph in Fig. 6.1

Node /Node	A	B	C	D	E	F	G	H	I
A	3	0	0	0	2	1	0	0	0
B	0	2	1	2	0	1	0	1	1
C	0	1	2	1	1	0	0	0	0
D	0	2	1	2	0	1	0	1	1
E	2	0	1	0	5	0	2	0	0
F	1	1	0	1	0	2	0	1	1
G	0	0	0	0	2	0	2	0	0
H	0	1	0	1	0	1	0	2	2
I	0	1	0	1	0	1	0	2	2

Table 6.3 Adjacency lists for the graph in Fig. 6.1

Node	Adjacency list
A	$\langle B, C, D \rangle$
B	$\langle A, E \rangle$
C	$\langle A, F \rangle$
D	$\langle A, E \rangle$
E	$\langle B, D, F, H, I \rangle$
F	$\langle C, E \rangle$
G	$\langle H, I \rangle$
H	$\langle E, G \rangle$
I	$\langle E, G \rangle$

- Note that some entries in A_G^2 are 0 indicating that there are no paths of length 2 between the corresponding pair of nodes. For example, the entry for node pair A, B is 0 meaning that there are no paths of length 2 between A and B.
- Observe that the entry for the pair A, E is 2 indicating that there are two paths of length 2. Further this value also characterizes the number of common neighbors between A and E.
- The number of paths of length 1 between A and F is 1. So, the number of common neighbors between A and F is 1; the common neighbor here is node C.
- **Adjacency Lists**:

   In the *adjacency lists* representation of a graph, we represent for each node the corresponding list of adjacent nodes. For the example graph in Fig. 6.1, the adjacency lists are given in Table 6.3.

   Adjacency lists representation offers sufficient *flexibility* to deal with *dynamic networks* where both the nodes and edges can get added or deleted over time. It is possible to obtain *common neighbors* of two nodes by considering nodes present in the intersection of the lists corresponding to the two nodes.

   However searching such sequential lists to find out whether a given edge/link is present or not can be *computationally prohibitive*.

   Such a list representation is popular in *information retrieval*.

   In the current treatment we deal only with adjacency matrix representation.

6.2   What Is a Social Network? [1–4]

Ideally a **social network** is a network where the nodes represent *humans* and the links/edges characterize interactions among humans. A typical and well-known example of such a network is exploited by *Facebook*. Here, *friendship* is the property characterized by a pair of nodes in the network and this property is *symmetric*.

It is convenient to extend the notion to entities other than humans as the resulting networks share a good number of interesting and useful properties. Some possibilities are as following:

6.2.1 Citation Networks

Here each node represents a paper P and an edge between a pair of nodes, P_i and P_j represents the fact that paper P_i cited paper P_j. Such a network has *directed* edges. So, a *directed graph* is used to represent it.

6.2.2 Coauthor Networks

Here each node corresponds to an author and there will be a link between two nodes if the corresponding authors have coauthored a paper. This can be represented by an undirected graph as *co-authorship* is symmetric.

6.2.3 Customer Networks

In these networks, the nodes characterize customers and there is a link between two customers if they have bought the same product in a departmental store.

6.2.4 Homogeneous and Heterogeneous Networks

In a *homogeneous network* all the nodes are of the same type. For example, a *friends network* is homogeneous. All the nodes in the network are humans and there is a link between two nodes if the corresponding people are friends.

On the other hand, in a *heterogeneous network* nodes could represent different types of entities. For example, in an academic network, it is possible to have both the authors and papers being represented by nodes. Further, the link between two authors could be based on *coauthorship* (A and B are coauthors), a link between an author and a paper could be *author of* (A is an author of P) relation, and a link between two papers is *cited* (P cited Q). So, the edges also could be of different types based on the types of their end vertices or nodes.

We consider analysis of only homogeneous social networks in this chapter.

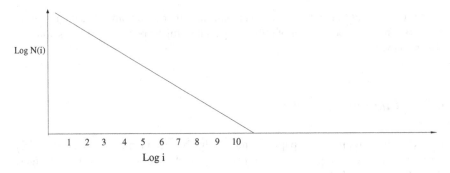

Fig. 6.2 Power-law degree distribution

6.3   Important Properties of Social Networks [4]

1. *Scale-Free Networks*: It is observed that *degree distribution* follows a *power law* asymptotically. Specifically, $N(i)$, number of nodes of degree i is given by

$$N(i) \propto i^{-\alpha}$$

   where α is found to be in the range from 2 to 3 based on empirical studies. It is called *scale-free* because the form of $N(i)$ does not change with different scales for i.
   A plot of the log $N(i)$ versus log i is shown in Fig. 6.2. Note that even if we scale i by some factor c, the values on the X axis will shift by a constant, that is $\log(c)$; this is because $\log(ci) = \log(c) + \log(i)$. Hence the specific form of the plot will not change.
   It means that in a given network there will be a large number of *low degree nodes* and a small number of *high degree nodes*. This property is exploited in the *analysis of social networks*.

2. **The Small-World Phenomenon**
   It is based on the observation that between any pair of nodes there is a short path made up of acquaintance/friendship links. This is also called as *six degrees of separation* where it was observed that the median path length between a pair of nodes is 6.
   This property is useful in the analysis that involves lengths of the paths between a pair of nodes.

3. **Homophily**
   One of the important properties of a social network is that *community structure* is manifest. So, nodes in a social network tend to form groups or clusters or more appropriately communities such that nodes in the same community are densely connected compared to nodes in different communities. The notion of homophily may be interpreted as a pair of *similar* nodes get connected.

Community structure is exploited in solving other important problems associated with social networks. This is because nodes in the same community are similar due to homophily.

6.4 Characterization of Communities [2, 3]

It is possible to represent a social network as a graph. In an abstract sense, a graph may be characterized using a *set of nodes* (V), a *set of edges/links* (E), and a *set of weights* (W). So,

- $G = \{V, E, W\}$, where
- V = set of nodes, $\{v_1, v_2, \ldots, v_n\}$
- E = set of edges $e_{i,j} \in E$ is the edge between v_i and v_j and
- W = set of weights

In a simple representation, we can have weight $w_{i,j} = 1$ if there is an edge between nodes v_i and v_j; if there is no edge, then $w_{i,j} = 0$. This corresponds to a *binary representation* that characterizes the *presence* or *absence* of an edge between pairs of nodes. It is possible to have a more general representation where $w_{i,j} \in \Re$; here $\Re$ is the set of real numbers. However, we deal with only binary representation in the rest of the chapter. In such a case we can simplify the notation and view the graph G as

- $G = \langle V, E \rangle$ where
- V = set of nodes, $\{v_1, v_2, \ldots, v_n\}$
- E = set of edges present; here edge $e_{i,j} \in E$ if there is an edge between nodes v_i and v_j, else $e_{i,j} \notin E$.

6.4.1 What Is a Community?

Intuitively we may say that nodes in a subset V^c of V are all in the same *community* if they are all similar to each other; or equivalently every pair of nodes in V^c are similar.

This is formally characterized using the notion of *clustering coefficient*, CC_i, defined as follows:

$$CC(v_i) = \frac{2 \mid E_{v_i} \mid}{\text{degree}_i \, (\text{degree}_i - 1)}$$

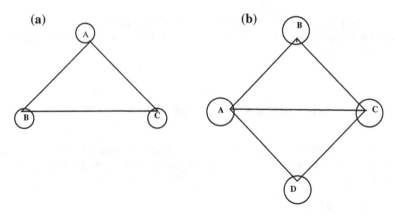

Fig. 6.3 Community strength based on connectedness. **a** Well connected nodes. **b** Not so well connected

- E_{v_i} is the set of edges among all pairs of neighbors of v_i.
- $|E_{v_i}|$ is the size of or number of elements in E_{v_i}.
- degree$_i$ is the degree of node v_i.

We illustrate this notion using the simple graphs given in Fig. 6.3.

In Fig. 6.3a, each node is connected to every other node as shown by the triangle. This is an example of a *clique of size 3*. Here, the clustering coefficient of each of the 3 nodes is 1. For example, consider Node A. Its neighbors are B and C and there is an edge between B and C. Note that

$|E_A| = 1$ and degree$_A = 2$.

So, $CC(A) = \frac{2|E_A|}{\text{degree}_A (\text{degree}_A - 1)} = 1$.

So, different nodes have the same CC value.

If we consider Fig. 6.3b, the pair of nodes B and D are not connected. So, the structure here is not a clique of size 4. So, clustering coefficient may not be 1 for all the nodes in this undirected graph. The values are:

1. $CC(A) = 0$ as B and D are neighbors of A and there is no edge between them.
2. $CC(B) = 1$ as there is a link between the neighbors of B, that is A and C.
3. Similar to A, $CC(C) = 0$ and
4. Similar to B, $CC(D) = 1$.

6.4.2   Clustering Coefficient of a Subgraph

The clustering coefficient of a cluster (or a subgraph) is the average of the clustering coefficients of all the nodes in the cluster (or the subgraph). Note that the clustering coefficient of the graph in Fig. 6.3a is 1, whereas that of the graph in Fig. 6.3b is 0.5.

It is possible to add meaningful edges/links between nodes that are not currently connected using *Link Prediction* that takes into account the similarity between the two end nodes of the possible link. This process increases the chances of the subgraphs to improve their connectivity or become cliques or near cliques so that the clusters as characterized by the clustering coefficient become more acceptable.

There are a variety of clustering algorithms that are used in detecting communities of entities in a network. These are graph-theoretic algorithms including

- Spectral clustering algorithms
- Single-link and complete-link algorithms
- Clustering based on Influential nodes in the graph

Community detection has several applications. For example, it will help in dealing with a community rather than individual nodes in classification based on *homophily*.

6.5 Link Prediction [1, 4]

Link prediction, in networks, is the process of predicting whether there will be a link between a pair of nodes that are not linked currently.

For example, consider the graph in Fig. 6.3b. There is no link between nodes B and D. Further, the nodes in the subgraph have a clustering coefficient of 0.5. If a link is added between B and D, then the clustering coefficient of the graph increases to 1.

There are several other important applications of link prediction.

- **Suggesting a paper** to be cited in a citation network. It is possible to use both the *structural properties* like the common neighbors of the two concerned paper nodes and also *semantic properties* like the common *keywords* between the two papers. Networks where both the structure and semantics are used are called *information networks*.
- **Recommending a Supervisor**: It involves recommending a supervisor B to a student A where A and B are interested in similar research areas.
- **Recommending a Collaborating Organization**: This involves suggesting a research group/university to an industry for possible collaboration.
- In heterogeneous networks, there could be several other applications like suggesting a book, a paper, a university, or a job to an individual.
- All these recommendations are based on the notion of similarity between a pair of nodes.

6.5.1 Similarity Between a Pair of Nodes

A pair of nodes are similar to each other if they have something in common either *structurally or semantically*. These are:

- **Structural Similarity**

  1. *Local similarity* This type of similarity is typically based on the *degree of each node* in the pair and/or the *common neighbors* of the two nodes. These similarity functions are based on the structure of the network or the graph representing it.
  2. *Global similarity* This kind of similarity is based on either the shortest path length between the two nodes or weighted number of paths between the two nodes. Here also we use graph structural properties.

- **Semantic Similarity**

  In this type of similarity, we consider the content associated with the nodes in the graph.

  1. *Keyword-based* A paper can cite another paper if both have a set of common keywords.
  2. *Fields of study* In Microsoft Academic Network, each paper or author is associated with a set of Fields of study. Two researchers may coauthor a paper if they have a good number of common fields of study.
  3. *Collaboration between two Organizations*

     It is possible to suggest an organization for possible collaboration to another organization based on common semantic properties like keywords and fields of study.

- **Dynamic and Static Networks**

  If a network evolves over time (or with respect to other parameters) then we say that the network is *dynamic*. In a dynamic network, the number of nodes and the number of edges can change over time.
  On the contrary, if a network does not change over time, then we call it a *static network*.
  Typically, most of the networks are dynamic; however, we can consider static snapshots of the network for possible analysis. In this chapter, we consider such static snapshots. Specifically, we assume that the set of nodes V is fixed and the set of edges E can evolve or change. In link prediction schemes we discuss here, we try to predict the possible additional links.

- **Local Similarity**

  Most of the popular local measures of similarity, between a pair of nodes, need to consider sets of neighbors of the nodes in the given undirected graph. So, we specify the notation first. Let

- $Ne(A) = \{x | A$ is linked with $x\}$ where $Ne(A)$ is the set of *neighbors of A*.
- $CN(A, B) == Ne(A) \cap Ne(B) =$ set of Common Neighbors of A and B.

We rank the links/edges to be added to the graph/network based on the similarity between the end vertices. So, we consider different ways the similarity could be specified. We consider some of the popular local similarity functions next.

6.6 Similarity Functions [1–4]

It is possible that either the network is *sparse* or *dense*. Typically, a dense network satisfies the power-law degree distribution, a sparse network may not satisfy. We consider functions that work well on sparse networks/graphs first. These are:

1. **Common Neighbors**: The similarity function is given by

   $cn(A, B) = | CN(A, B) | =$ Number of Common Neighbors of A and B.

   This captures the notion that larger the number of common friends of two people better the possibility of the two people getting connected. It does not consider the degrees of the common neighbor nodes.

2. **Jaccard's Coefficient**: The Jaccard's coefficient, jc, may be viewed as a normalized version of cn. It is given by

   $$jc(A, B) = \frac{| Ne(A) \cap Ne(B) |}{| Ne(A) \cup Ne(B) |} = \frac{cn(A, B)}{| Ne(A) \cup Ne(B) |}.$$

   It uses the size of the union of sets of neighbors of A and B to normalize the cn score.

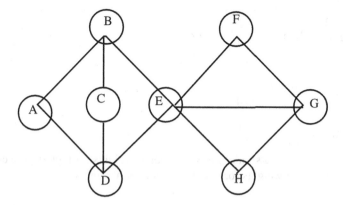

Fig. 6.4 Local similarity functions

6.6.1 Example

We illustrate these local similarity functions using the example graph shown in Fig. 6.4.

1. *Common Neighbors*
   $cn(B, D) = | \{A, C, E\} | = 3$
   $cn(A, E) = 2, cn(A, G) = 0.$
2. *Jaccard's Coefficient*
   $jc(B, D) = \frac{|\{A,C,E\}|}{|\{A,C,E\}|} = 1$

   $jc(A, G) = 0$

In the case of *dense networks*, the network satisfies power-law degree distribution. In such a case, one can exploit the degree information of the common neighbors in getting a better similarity value between a pair of nodes. Two popular *local measures* for dense networks are

1. *Adamic-Adar*: Here the similarity is a weighted version of the common neighbors where the weight is inversely proportional to the logarithm of the degree of the common neighbor. The adamic-adar, *aa* similarity is defined as

$$aa(A, B) = \sum_{v_i \in CN(A,B)} \frac{1}{\log | Ne(v_i) |}$$

2. *Resource Allocation Index*
   The resource allocation (*ra*) similarity index is a minor variant of *aa* where the weight of a common neighbor is inversely proportional to the degree of the common neighbor, instead of logarithm of the degree.

We illustrate these similarity functions using the graph shown in Fig. 6.4.

1. *Adamic-Adar*
   $aa(B, D) = \frac{1}{\log 2} + \frac{1}{\log 2} + \frac{1}{\log 5} = 2.4$

   $aa(A, G) = 0$
2. *Resource Allocation*
   $ra(B, D) = \frac{1}{2} + \frac{1}{2} + \frac{1}{5} = 1.2$

   $ra((A, G) = 0$

Note that both *aa* and *ra* similarities give smaller weights for high degree common neighbors and larger weights for low degree common neighbors.

6.6.2 Global Similarity

Global similarity between a pair of nodes will be based on a global computation. Note that either the *cn* or *jc* value between A and G is 0 because these similarity values are based on the *local structure* around the nodes in the pair. However, global similarity could be computed between a pair of nodes that may not share any local structure. Again, for sparse networks, the similarity is based on the degree of the end vertices.

- **Preferential Attachment**: Here, the similarity $pa(A, B)$ is given by

$$pa(A, B) = | Ne(A) | \times | Ne(B) | .$$

  This function prefers edges between a pair of high degree nodes; this makes sense when the graph is sparse.
- *Example in Fig.* 6.4
  Note that $pa(A, E) = 2 \times 5 = 10$
  $pa(A, C) = pa(F, H) = 4$ and $pa(A, G) = 6$.

In the case of dense networks, the global similarity functions exploit the distances and/or paths between the two nodes. Two popular functions are:

1. **Graph Distance**: Here, the similarity, *gds*, between a pair of nodes A and B is inversely proportional to the length of the shortest path between A and B.

$$gds(A, B) = \frac{1}{\text{length of the shortest path}(A, B)}.$$

2. **Katz Similarity** (*ks*): It is based on number of paths of some length l and each such number is weighted based on a function of l. The weights are such that shorter paths get higher weights and longer paths get smaller weights. Specifically, it is given by

$$ks(A, B) = \sum_{l=1}^{\infty} \beta^l . | npath_l | .$$

  where $npath_l$ is the number of paths of length l between A and B.

We illustrate these similarity functions using the graph in Fig. 6.4.

1. $gds(A, C) = \frac{1}{2} = 0.5$. Note that there are two shortest paths between A and C; one is through B and the other is via node D. Both are of length 2.
   $gds(A, G) = 0.33$
2. $ks(A, C) = 0.02$ and $ks(A, G) = 0.00262$. Note that there are two paths of length 3; two paths of length 5, and six paths of length 4 between A and G. Further, the value of β is assumed to be 0.1.

Community	SVM prediction accuracy (%)
C_1	85.33
C_2	85
C_3	67
C_4	80
C_5	80

Table 6.4 Link prediction based on linear SVM

- *Structural and Semantic Properties for Link Prediction*
  It is possible to combine structure and semantics to extract features and use them in supervised learning. A simple *binary classifier* that will learn two classes

  - **Positive class**: a link exists between the pair of nodes
  - **Negative class**: there is no link between the two nodes
  - **Features**: both structural and semantic features. For example, in a citation network
    1. **Structural** local and global similarity values between the pair of nodes
    2. **Semantic** keywords from the papers corresponding to the two nodes and also from their neighboring nodes.

6.6.3 Link Prediction based on Supervised Learning

We conducted an experiment to test how link prediction can be done using SVM classifier. The algorithm was specified by Hasan et al. details of which can be found in the reference given at the end of the chapter.

For the sake of experimentation, we synthesized a network having 100 nodes. We randomly formed five communities having 20 nodes each, where the density of links is higher when compared to the rest of the network. Further, this data is divided into train and test sets based on the edges. We choose pairs of nodes present in the train set with no links between them, but are connected in the test set as positive patterns and the ones which are not connected in the test set as negative patterns. We learn a Linear SVM for predicting the links in the test set, or equivalently to build a binary classifier and use it in classification.

We observed that the link prediction based on SVM classifier works well within the communities rather than across communities in the network. Further, we have different number of positive and negative patterns for each community. The community C_3 has more class imbalance. The SVM prediction accuracy results for the five communities are given in Table 6.4.

6.7 Summary

1. Networks are playing an important role in several applications.
2. Friends network maintained by Facebook, Twitter network supported by Twitter and academic network supported by Microsoft are some examples of well-studied networks.
3. There are several properties that are satisfied by most of the networks. These properties include *power-law degree distribution, six degrees of separation* on an average between a pair of nodes, and *exhibiting community structure*. These properties are useful in analyzing social networks.
4. We have examined *link prediction* in more detail here.
5. We have considered the role of both *local* and *global* similarity measures.
6. We have described the role of *supervised learning* in link prediction which can exploit both structural and semantic properties of nodes in the network.
7. It is not only possible to analyze social networks where nodes are humans but also other networks like citation networks and term cooccurence networks. Several such networks can be analyzed using a set of properties that are satisfied by networks.

References

1. Hasan, M.A., Chaoji, V., Salem, S., Zaki, M.: Link prediction using supervised learning. In: Proceedings of SDM 06 Workshop on Counter Terrorism and Security, 20–22 April, 2006, Bethesda, Maryland, USA (2006)
2. Leskovec, J., Rajaraman, A., Ullman, J.: Mining oF Massive Datasets, Cambridge University Press (2014)
3. Liben-Nowell, D., Kleinberg, J.M.: The link prediction problem for social networks. In: Proceedings of CIKM, 03–08 Nov 2003, New Orleans, LA, USA, pp. 556–559 (2003)
4. Virinchi, S., Mitra, P.: Link Prediction in Social Networks: Role of Power Law Distribution. Springer, Springer Briefs in Computer Science (2016)

Chapter 7
Conclusion

Abstract In this chapter, we conclude by looking at various properties of linear classifiers, piecewise linear classifiers, and nonlinear classifiers. We look at the issues of learning and optimization associated with linear classifiers.

Keywords Perceptron · SVM · Optimization · Learning · Kernel trick · Supervised link prediction

In this book we have examined some of the well-known linear classifiers. Specifically, we considered classifiers based on linear discriminant functions. Some of the specific features of the book are:

1. We have discussed three types of classifiers

    a. *Linear classifiers*
    b. *Piecewise linear classifiers*
    c. *Nonlinear classifiers.*

2. We have discussed classifiers like *NNC* and *KNNC* which are inherently *nonlinear*. We have indicated how the discriminant function framework can be used to characterize these classifiers.
3. We have discussed on how a *piecewise linear* classifier like the *DTC* can be characterized using discriminant functions that are based on logical expressions.
4. Also we have considered well-known linear classifiers like the *MDC* and the *Minimal Mahalanobis Distance Classifier* that are inherently linear in a two-class setting. They can be optimal under some conditions on the data distributions.
5. Two popularly used classifiers in *text mining* are *SVM* and *NBC*. We have indicated the inherent linear structure of the *NBC* in a two-class setting.
6. It is possible to represent even *nonlinear discriminant functions* in the form of linear discriminant function, possibly in a higher dimensional space. If we know the nonlinear form explicitly, then we can directly convert it into a linear function and use all possible linear classifiers.
7. We described *linear discriminant functions* and show how they can be used in linear classification.

M.N. Murty and R. Raghava, *Support Vector Machines and Perceptrons*,
SpringerBriefs in Computer Science, DOI 10.1007/978-3-319-41063-0_7

8. We indicated how the *weight vector W* and *threshold b* characterize the *decision boundary* between the two classes and the role of *negative* and *positive half spaces* in binary classification.

9. We described classification using *perceptron*. We dealt with the *perceptron learning algorithm* and its *convergence* in the *linearly separable* case.

10. We have justified the *weight update rule* using *algebraic, geometric, and optimization* viewpoints.

11. We have indicated how the perceptron weight vector can be viewed as a *weighted combination of the training patterns*. These weights are based on the class label and the number of times a pattern is misclassified by the weight vectors in the earlier iterations.

12. The most *important theoretical foundation* of perceptrons was provided by *Minsky and Papert* in their book on *Perceptrons*. This deals with the notion of *order of a perceptron*. They say that for some simple predicates the order could be 1 and hence it is easy to compute the predicate in a distributed and/or incremental manner. However, for predicates like the *exclusive or* the order keeps increasing as we increase the number of boolean/binary variables; so computation is more difficult. The associated theoretical notions like *order of a perceptron, permutation invariance, positive normal form* that uses *minterms* are discussed in a simple manner through suitable examples.

13. We have discussed some similarities and differences between SVMs and perceptrons.

14. We explained the notions of *margin, hard margin formulation,* and *soft margin formulation* associated with the linear SVM that maximizes the margin under some constraints. This is a well-behaved *convex optimization* problem that offers a globally optimum solution.

15. We explained how *multiclass problems* can be solved using a *combination of binary classifiers*.

16. We discussed the *kernel trick* that can be exploited in dealing with nonlinear discriminant functions using linear discriminant functions in high-dimensional spaces.

17. The kernel trick could be used to perform computations in the low-dimensional input space instead of a possibly infinite dimensional feature or kernel space.

18. The theory of kernel functions permits dealing with possible infinite dimensional spaces which could be realized using exponential kernel functions. Such functions will have infinite terms in their series expansions. However, the theory behind perceptrons considers all possible computable functions based on boolean representations; in such a boolean representation that is natural to a digital computer, there is no scope for infinite dimensions. In a perceptron, we may need to use all possible minterms that could be formed using some d boolean features and the number of such minterms will never be more than 2^d.

19. We have compared the performance of perceptrons and SVMs on some practical datasets.

20. We have considered the application of SVMs in *link prediction* in social networks. We briefly discussed *social networks*, their *important properties*, and several types of techniques dealing with such networks. Specifically, we have examined

    a. Community detection and clustering coefficient
    b. Link prediction using local and global similarity measures
    c. The role of SVM in link prediction based on supervised learning.

Glossary

$g(X)$	Linear Discriminant Function
μ	Mean of a class
Σ	Covariance Matrix
C_+	Positive class
C_-	Negative class
W	Weight vector
b	Threshold weight
α	Weight of support vector
$\mathcal{L}$	Lagrangian
X_i	ith pattern
y_i	Class label of the ith pattern
G	Graph representing a network
V	Set of vertices or nodes in the graph
E	Set of edges in a graph

© The Author(s) 2016

M.N. Murty and R. Raghava, *Support Vector Machines and Perceptrons*,
SpringerBriefs in Computer Science, DOI 10.1007/978-3-319-41063-0

Index

© The Author(s) 2016
M.N. Murty and R. Raghava, *Support Vector Machines and Perceptrons*,
SpringerBriefs in Computer Science, DOI 10.1007/978-3-319-41063-0

Printed in the United States
By Bookmasters